Data Mining Techniques in CRM

Data Mining Techniques in CRM

Inside Customer Segmentation

Konstantinos Tsiptsis

CRM & Customer Intelligence Expert, Athens, Greece

Antonios Chorianopoulos

Data Mining Expert, Athens, Greece

A John Wiley and Sons, Ltd., Publication

This edition first published 2009
© 2009, John Wiley & Sons, Ltd

Registered office

John Wiley & Sons Ltd, The Atrium, Southern Gate, Chichester, West Sussex, PO19 8SQ, United Kingdom

For details of our global editorial offices, for customer services and for information about how to apply for permission to reuse the copyright material in this book please see our website at www.wiley.com.

The right of the author to be identified as the author of this work has been asserted in accordance with the Copyright, Designs and Patents Act 1988.

Reprinted in 2009, September 2011

Library of Congress Cataloguing-in-Publication Data

Record on file

A catalogue record for this book is available from the British Library.

ISBN: 978-0-470-74397-3 (H/B)

Typeset in 11/13.5pt NewCaledonia by Laserwords Private Limited, Chennai, India.
Printed and bound by CPI Group (UK) Ltd, Croydon, CR0 4YY

To my daughter Eugenia and my wife Virginia, for their support and understanding.
And to my parents.
– Antonios

In memory of my father.
Dedicated to my daughters Marcella and Christina, my wife Maria, my sister Marina and
my niece Julia and of course, to my mother Maria who taught me to set my goals in life.
– Konstantinos

CONTENTS

ACKNOWLEDGEMENTS   **xiii**

1 DATA MINING IN CRM   **1**
   The CRM Strategy   1
   What Can Data Mining Do?   2
      Supervised/Predictive Models   3
      Unsupervised Models   3
      Data Mining in the CRM Framework   4
         Customer Segmentation   4
         Direct Marketing Campaigns   5
         Market Basket and Sequence Analysis   7
      The Next Best Activity Strategy and "Individualized" Customer Management   8
   The Data Mining Methodology   10
   Data Mining and Business Domain Expertise   13
   Summary   13

2 AN OVERVIEW OF DATA MINING TECHNIQUES   **17**
   Supervised Modeling   17
      Predicting Events with Classification Modeling   19
         Evaluation of Classification Models   25
         Scoring with Classification Models   32
      Marketing Applications Supported by Classification Modeling   32
      Setting Up a Voluntary Churn Model   33
      Finding Useful Predictors with Supervised Field Screening Models   36
      Predicting Continuous Outcomes with Estimation Modeling   37
   Unsupervised Modeling Techniques   39
      Segmenting Customers with Clustering Techniques   40
      Reducing the Dimensionality of Data with Data Reduction Techniques   47
      *Finding "What Goes with What" with Association or Affinity Modeling
         Techniques*   50
      Discovering Event Sequences with Sequence Modeling Techniques   56
      Detecting Unusual Records with Record Screening Modeling Techniques   59
   Machine Learning/Artificial Intelligence vs. Statistical Techniques   61
   Summary   62

3   DATA MINING TECHNIQUES FOR SEGMENTATION   **65**

Segmenting Customers with Data Mining Techniques   65
Principal Components Analysis   65
   PCA Data Considerations   67
   How Many Components Are to Be Extracted?   67
   What Is the Meaning of Each Component?   75
   Does the Solution Account for All the Original Fields? 78
   Proceeding to the Next Steps with the Component Scores 79
   Recommended PCA Options   80
Clustering Techniques   82
   Data Considerations for Clustering Models   83
   Clustering with K-means   85
     Recommended K-means Options   87
   Clustering with the TwoStep Algorithm   88
     Recommended TwoStep Options   90
   Clustering with Kohonen Network/Self-organizing Map   91
     Recommended Kohonen Network/SOM Options   93
Examining and Evaluating the Cluster Solution   96
   The Number of Clusters and the Size of Each Cluster   96
   Cohesion of the Clusters   97
   Separation of the Clusters   99
Understanding the Clusters through Profiling   100
   Profiling the Clusters with IBM SPSS Modeler's Cluster Viewer 102
   Additional Profiling Suggestions   105
Selecting the Optimal Cluster Solution   108
Cluster Profiling and Scoring with Supervised Models   110
An Introduction to Decision Tree Models   110
   The Advantages of Using Decision Trees for Classification Modeling 121
   One Goal, Different Decision Tree Algorithms: C&RT, C5.0, and CHAID 123
     Recommended CHAID Options   125
Summary   127

4   THE MINING DATA MART   **133**

Designing the Mining Data Mart   133
The Time Frame Covered by the Mining Data Mart   135
The Mining Data Mart for Retail Banking   137
   Current Information   138
     Customer Information   138
     Product Status   138
   Monthly Information   140
     Segment and Group Membership   141
     Product Ownership and Utilization   141
     Bank Transactions   141
   Lookup Information   143
     Product Codes   144

Transaction Channels	145
Transaction Types	145
The Customer "Signature" – from the Mining Data Mart to the Marketing	
Customer Information File	148
Creating the MCIF through Data Processing	149
Derived Measures Used to Provide an "Enriched" Customer View	154
The MCIF for Retail Banking	155
The Mining Data Mart for Mobile Telephony Consumer (Residential) Customers	160
Mobile Telephony Data and CDRs	162
Transforming CDR Data into Marketing Information	162
Current Information	163
Customer Information	164
Rate Plan History	165
Monthly Information	167
Outgoing Usage	167
Incoming Usage	169
Outgoing Network	170
Incoming Network	170
Lookup Information	170
Rate Plans	171
Service Types	171
Networks	172
The MCIF for Mobile Telephony	172
The Mining Data Mart for Retailers	177
Transaction Records	179
Current Information	179
Customer Information	179
Monthly Information	180
Transactions	180
Purchases by Product Groups	182
Lookup Information	183
The Product Hierarchy	183
The MCIF for Retailers	184
Summary	187
5 CUSTOMER SEGMENTATION	**189**
An Introduction to Customer Segmentation	189
Segmentation in Marketing	190
Segmentation Tasks and Criteria	191
Segmentation Types in Consumer Markets	191
Value-Based Segmentation	193
Behavioral Segmentation	194
Propensity-Based Segmentation	195
Loyalty Segmentation	196
Socio-demographic and Life-Stage Segmentation	198

Needs/Attitudinal-Based Segmentation	199
Segmentation in Business Markets	200
A Guide for Behavioral Segmentation	203
Behavioral Segmentation Methodology	203
Business Understanding and Design of the Segmentation Process	203
Data Understanding, Preparation, and Enrichment	205
Identification of the Segments with Cluster Modeling	208
Evaluation and Profiling of the Revealed Segments	208
Deployment of the Segmentation Solution, Design, and Delivery of Differentiated Strategies	211
Tips and Tricks	211
Segmentation Management Strategy	213
A Guide for Value-Based Segmentation	216
Value-Based Segmentation Methodology	216
Business Understanding and Design of the Segmentation Process	216
Data Understanding and Preparation –Calculation of the Value Measure	218
Grouping Customers According to Their Value	218
Profiling and Evaluation of the Value Segments	219
Deployment of the Segmentation Solution	219
Designing Differentiated Strategies for the Value Segments	220
Summary	223
6 SEGMENTATION APPLICATIONS IN BANKING	**225**
Segmentation for Credit Card Holders	225
Designing the Behavioral Segmentation Project	226
Building the Mining Dataset	227
Selecting the Segmentation Population	228
The Segmentation Fields	230
The Analytical Process	233
Revealing the Segmentation Dimensions	233
Identification and Profiling of Segments	237
Using the Segmentation Results	256
Behavioral Segmentation Revisited: Segmentation According to All Aspects of Card Usage	258
The Credit Card Case Study: A Summary	263
Segmentation in Retail Banking	264
Why Segmentation?	264
Segmenting Customers According to Their Value: The Vital Few Customers	267
Using Business Rules to Define the Core Segments	268
Segmentation Using Behavioral Attributes	271
Selecting the Segmentation Fields	271
The Analytical Process	274
Identifying the Segmentation Dimensions with PCA/Factor Analysis	274
Segmenting the "Pure Mass" Customers with Cluster Analysis	276
Profiling of Segments	276

The Marketing Process 283
    Setting the Business Objectives 283
Segmentation in Retail Banking: A Summary 288

7 SEGMENTATION APPLICATIONS IN TELECOMMUNICATIONS **291**
Mobile Telephony 291
Mobile Telephony Core Segments – Selecting the Segmentation Population 292
Behavioral and Value-Based Segmentation – Setting Up the Project 294
Segmentation Fields 295
Value-Based Segmentation 300
Value-Based Segments: Exploration and Marketing Usage 304
Preparing Data for Clustering – Combining Fields into Data Components 307
Identifying, Interpreting, and Using Segments 313
Segmentation Deployment 326
The Fixed Telephony Case 329
Summary 331

8 SEGMENTATION FOR RETAILERS **333**
Segmentation in the Retail Industry 333
The RFM Analysis 334
The RFM Segmentation Procedure 338
RFM: Benefits, Usage, and Limitations 345
Grouping Customers According to the Products They Buy 346
Summary 348

FURTHER READING **349**

INDEX **351**

ACKNOWLEDGEMENTS

We would like to thank Vlassis Papapanagis, Leonidas Georgiou and Ioanna Koutrouvis of SPSS, Greece. Also, Andreas Kokkinos, George Krassadakis, Kyriakos Kokkalas and Loukas Maragos. Special thanks to Ioannis Mataragas for the creation of the line drawings.

CHAPTER ONE

Data Mining in CRM

THE CRM STRATEGY

Customers are the most important asset of an organization. There cannot be any business prospects without satisfied customers who remain loyal and develop their relationship with the organization. That is why an organization should plan and employ a clear strategy for treating customers. CRM (Customer Relationship Management) is the strategy for building, managing, and strengthening loyal and long-lasting customer relationships. CRM should be a customer-centric approach based on customer insight. Its scope should be the "personalized" handling of customers as distinct entities through the identification and understanding of their differentiated needs, preferences, and behaviors.

In order to make the CRM objectives and benefits clearer, let us consider the following real-life example of two clothing stores with different selling approaches. Employees of the first store try to sell everything to everyone. In the second store, employees try to identify each customer's needs and wants and make appropriate suggestions. Which store will finally look more reliable in the eyes of customers? Certainly the second one seems more trustworthy for a long-term relationship, since it aims for customer satisfaction by taking into account the specific customer needs.

CRM has two main objectives:

1. Customer retention through customer satisfaction.
2. Customer development through customer insight.

The importance of the first objective is obvious. Customer acquisition is tough, especially in mature markets. It is always difficult to replace existing customers

Data Mining Techniques in CRM: Inside Customer Segmentation   K. Tsiptsis and A. Chorianopoulos
© 2009 John Wiley & Sons, Ltd

with new ones from the competition. With respect to the second CRM goal of customer development, the key message is that there is no average customer. The customer base comprises different persons, with different needs, behaviors, and potentials that should be handled accordingly.

Several CRM software packages are available and used to track and efficiently organize inbound and outbound interactions with customers, including the management of marketing campaigns and call centers. These systems, referred to as operational CRM systems, typically support front-line processes in sales, marketing, and customer service, automating communications and interactions with the customers. They record contact history and store valuable customer information. They also ensure that a consistent picture of the customer's relationship with the organization is available at all customer "touch" (interaction) points.

However, these systems are just tools that should be used to support the strategy of effectively managing customers. To succeed with CRM and address the aforementioned objectives, organizations need to gain insight into customers, their needs, and wants through data analysis. This is where analytical CRM comes in. Analytical CRM is about analyzing customer information to better address the CRM objectives and deliver the right message to the right customer. It involves the use of data mining models in order to assess the value of the customers, understand, and predict their behavior. It is about analyzing data patterns to extract knowledge for optimizing the customer relationships.

For example, data mining can help in customer retention as it enables the timely identification of valuable customers with increased likelihood to leave, allowing time for targeted retention campaigns. It can support customer development by matching products with customers and better targeting of product promotion campaigns. It can also help to reveal distinct customer segments, facilitating the development of customized new products and product offerings which better address the specific preferences and priorities of the customers.

The results of the analytical CRM procedures should be loaded and integrated into the operational CRM front-line systems so that all customer interactions can be more effectively handled on a more informed and "personalized" base. This book is about analytical CRM. Its scope is to present the application of data mining techniques in the CRM framework and it especially focuses on the topic of customer segmentation.

WHAT CAN DATA MINING DO?

Data mining aims to extract knowledge and insight through the analysis of large amounts of data using sophisticated modeling techniques. It converts data into knowledge and actionable information.

The data to be analyzed may reside in well-organized data marts and data warehouses or may be extracted from various unstructured data sources. A data mining procedure has many stages. It typically involves extensive data management before the application of a statistical or machine learning algorithm and the development of an appropriate model. Specialized software packages have been developed (data mining tools), which can support the whole data mining procedure.

Data mining models consist of a set of rules, equations, or complex "transfer functions" that can be used to identify useful data patterns, understand, and predict behaviors. They can be grouped into two main classes according to their goal, as follows.

SUPERVISED/PREDICTIVE MODELS

In supervised, or predictive, directed, or targeted modeling, the goal is to predict an event or estimate the values of a continuous numeric attribute. In these models there are input fields or attributes and an output or target field. Input fields are also called predictors because they are used by the model to identify a prediction function for the output field. We can think of predictors as the X part of the function and the target field as the Y part, the outcome.

The model uses the input fields which are analyzed with respect to their effect on the target field. Pattern recognition is "supervised" by the target field. Relationships are established between input and output fields. An input–output mapping "function" is generated by the model, which associates predictors with the output and permits the prediction of the output values, given the values of the input fields.

Predictive models are further categorized into classification and estimation models:

- **Classification or propensity models:** In these models the target groups or classes are known from the start. The goal is to classify the cases into these predefined groups; in other words, to predict an event. The generated model can be used as a scoring engine for assigning new cases to the predefined classes. It also estimates a propensity score for each case. The propensity score denotes the likelihood of occurrence of the target group or event.
- **Estimation models:** These models are similar to classification models but with one major difference. They are used to predict the value of a continuous field based on the observed values of the input attributes.

UNSUPERVISED MODELS

In unsupervised or undirected models there is no output field, just inputs. The pattern recognition is undirected; it is not guided by a specific target attribute.

The goal of such models is to uncover data patterns in the set of input fields. Unsupervised models include:

- **Cluster models:** In these models the groups are not known in advance. Instead we want the algorithms to analyze the input data patterns and identify the natural groupings of records or cases. When new cases are scored by the generated cluster model they are assigned to one of the revealed clusters.
- **Association and sequence models:** These models also belong to the class of unsupervised modeling. They do not involve direct prediction of a single field. In fact, all the fields involved have a double role, since they act as inputs and outputs at the same time. Association models detect associations between discrete events, products, or attributes. Sequence models detect associations over time.

DATA MINING IN THE CRM FRAMEWORK

Data mining can provide customer insight, which is vital for establishing an effective CRM strategy. It can lead to personalized interactions with customers and hence increased satisfaction and profitable customer relationships through data analysis. It can support an 'individualized' and optimized customer management throughout all the phases of the customer lifecycle, from the acquisition and establishment of a strong relationship to the prevention of attrition and the winning back of lost customers. Marketers strive to get a greater market share and a greater share of their customers. In plain words, they are responsible for getting, developing, and keeping the customers. Data mining models can help in all these tasks, as shown in Figure 1.1.

More specifically, the marketing activities that can be supported with the use of data mining include the following topics.

Customer Segmentation

Segmentation is the process of dividing the customer base into distinct and internally homogeneous groups in order to develop differentiated marketing strategies according to their characteristics. There are many different segmentation types based on the specific criteria or attributes used for segmentation.

In behavioral segmentation, customers are grouped by behavioral and usage characteristics. Although behavioral segments can be created with business rules, this approach has inherent disadvantages. It can efficiently handle only a few segmentation fields and its objectivity is questionable as it is based on the personal perceptions of a business expert. Data mining on the other hand can create

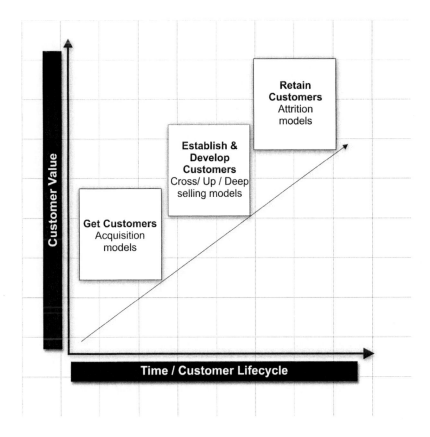

Figure 1.1   Data mining and customer lifecycle management.

data-driven behavioral segments. Clustering algorithms can analyze behavioral data, identify the natural groupings of customers, and suggest a solution founded on observed data patterns. Provided the data mining models are properly built, they can uncover groups with distinct profiles and characteristics and lead to rich segmentation schemes with business meaning and value.

Data mining can also be used for the development of segmentation schemes based on the current or expected/estimated value of the customers. These segments are necessary in order to prioritize customer handling and marketing interventions according to the importance of each customer.

Direct Marketing Campaigns

Marketers use direct marketing campaigns to communicate a message to their customers through mail, the Internet, e-mail, telemarketing (phone), and other direct channels in order to prevent churn (attrition) and to drive customer acqui-sition and purchase of add-on products. More specifically, acquisition campaigns aim at drawing new and potentially valuable customers away from the competition. Cross-/deep-/up-selling campaigns are implemented to sell additional products, more of the same product, or alternative but more profitable products to existing customers. Finally, retention campaigns aim at preventing valuable customers from terminating their relationship with the organization.

When not refined, these campaigns, although potentially effective, can also lead to a huge waste of resources and to bombarding and annoying customers with unsolicited communications. Data mining and classification (propensity) models in particular can support the development of targeted marketing campaigns. They analyze customer characteristics and recognize the profiles of the target customers. New cases with similar profiles are then identified, assigned a high propensity score, and included in the target lists. The following classification models are used to optimize the subsequent marketing campaigns:

- **Acquisition models:** These can be used to recognize potentially profitable prospective customers by finding "clones" of valuable existing customers in external lists of contacts,
- **Cross-/deep-/up-selling models:** These can reveal the purchasing potential of existing customers.
- **Voluntary attrition or voluntary churn models:** These identify early churn signals and spot those customers with an increased likelihood to leave volun-tarily.

When properly built, these models can identify the right customers to contact and lead to campaign lists with increased density/frequency of target customers. They outperform random selections as well as predictions based on business rules and personal intuition. In predictive modeling, the measure that compares the predictive ability of a model to randomness is called the lift. It denotes how much better a classification data mining model performs in comparison to a random selection. The "lift" concept is illustrated in Figure 1.2 which compares the results of a data mining churn model to random selection.

In this hypothetical example, a randomly selected sample contains 10% of actual "churners." On the other hand, a list of the same size generated by a data mining model is far more effective since it contains about 60% of actual churners. Thus, data mining achieved six times better predictive ability than randomness. Although completely hypothetical, these results are not far from reality. Lift values higher than 4, 5, or even 6 are quite common in those real-world situations that

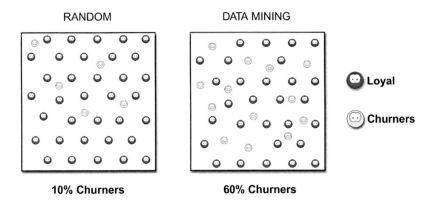

Figure 1.2 The increase in predictive ability resulting from the use of a data mining churn model.

were appropriately tackled by well-designed propensity models, indicating the potential for improvement offered by data mining.

The stages of direct marketing campaigns are illustrated in Figure 1.3 and explained below:

1. Gathering and integrating the necessary data from different data sources.
2. Customer analysis and segmentation into distinct customer groups.
3. Development of targeted marketing campaigns by using propensity models in order to select the right customers.
4. Campaign execution by choosing the appropriate channel, the appropriate time, and the appropriate offer for each campaign.
5. Campaign evaluation through the use of test and control groups. The evaluation involves the partition of the population into test and control groups and comparison of the positive responses.
6. Analysis of campaign results in order to improve the campaign for the next round in terms of targeting, time, offer, product, communication, and so on.

Data mining can play a significant role in all these stages, particularly in identifying the right customers to be contacted.

Market Basket and Sequence Analysis

Data mining and association models in particular can be used to identify related products typically purchased together. These models can be used for market basket analysis and for revealing bundles of products or services that can be sold together.

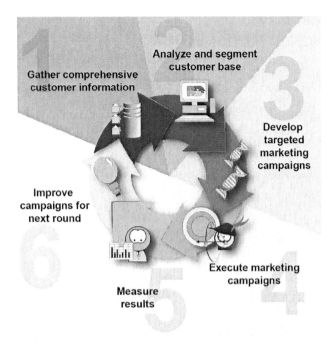

Figure 1.3   The stages of direct marketing campaigns.

Sequence models take into account the order of actions/purchases and can identify sequences of events.

THE NEXT BEST ACTIVITY STRATEGY AND "INDIVIDUALIZED" CUSTOMER MANAGEMENT

The data mining models should be put together and used in the everyday business operations of an organization to achieve more effective customer management. The knowledge extracted by data mining can contribute to the design of a next best activity (NBA) strategy. More specifically, the customer insight gained by data mining can enable the setting of "personalized" marketing objectives. The organization can decide on a more informed base the next best marketing activity for each customer and select an "individualized" approach which might be the following:

- An offer for preventing attrition, mainly for high-value, at-risk customers.
- A promotion for the right add-on product and a targeted cross-/up-/deep-selling offer for customers with growth potential.

- Imposing usage limitations and restrictions on customers with bad payment records and bad credit risk scores.
- The development of a new product/offering tailored to the specific characteristics of an identified segment, and so on.

The main components that should be taken into account in the design of the NBA strategy are illustrated in Figure 1.4. They are:

1. The current and expected/estimated customer profitability and value.
2. The type of customer, the differentiating behavioral and demographic characteristics, the identified needs and attitudes revealed through data analysis and segmentation.
3. The growth potential as designated by relevant cross-/up-/deep-selling models and propensities.

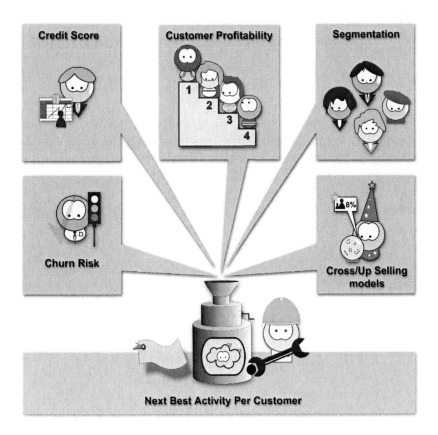

Figure 1.4   The next best activity components.

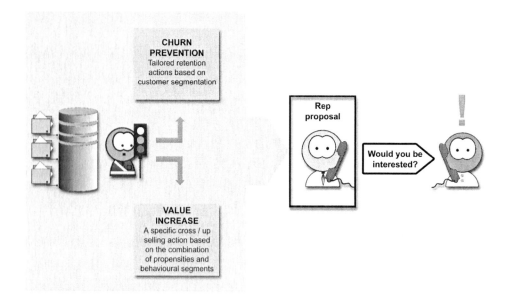

Figure 1.5   The next best activity strategy in action.

4. The defection risk/churn propensity as estimated by a voluntary churn model.
5. The payment behavior and credit score of the customer.

In order to better understand the role of these components and see the NBA strategy in action (Figure 1.5), let us consider the following simple example. A high-value banking customer has a high potential of getting a mortgage loan but at the same time is also scored with a high probability to churn. What is the best approach for this customer and how should he be handled by the organization? As a high-value, at-risk customer, the top priority is to prevent his leaving and lure him with an offer that matches his particular profile. Therefore, instead of receiving a cross-selling offer, he should be included in a retention campaign and contacted with an offer tailored to the specific characteristics of the segment to which he belongs.

THE DATA MINING METHODOLOGY

A data mining project involves more than modeling. The modeling phase is just one phase in the implementation process of a data mining project. Steps of critical importance precede and follow model building and have a significant effect on the success of the project.

Table 1.1   The CRISP-DM phases.

1. Business understanding	2. Data understanding	3. Data preparation
• Understanding the business goal • Situation assessment • Translating the business goal into a data mining objective • Development of a project plan	• Considering data requirements • Initial data collection, exploration, and quality assessment	• Selection of required data • Data acquisition • Data integration and formatting (merge/joins, aggregations) • Data cleaning • Data transformations and enrichment (regrouping/binning of existing fields, creation of derived attributes and key performance indicators: ratios, flag fields, averages, sums, etc.)
4. Modeling	**5. Model evaluation**	**6. Deployment**
• Selection of the appropriate modeling technique • Especially in the case of predictive models, splitting of the dataset into training and testing subsets for evaluation purposes • Development and examination of alternative modeling algorithms and parameter settings • Fine tuning of the model settings according to an initial assessment of the model's performance	• Evaluation of the model in the context of the business success criteria • Model approval	• Create a report of findings • Planning and development of the deployment procedure • Deployment of the data mining model • Distribution of the model results and integration in the organization's operational CRM system • Development of a maintenance–update plan • Review of the project • Planning the next steps

An outline of the basic phases in the development of a data mining project according to the CRISP-DM (Cross Industry Standard Process for Data Mining) process model is presented in Table 1.1.

Data mining projects are not simple. They usually start with high expectations but may end in business failure if the engaged team is not guided by a clear methodological framework. The CRISP-DM process model charts the steps that should be followed for successful data mining implementations. These steps are as follows:

1. **Business understanding:** The data mining project should start with an understanding of the business objective and an assessment of the current situation. The project's parameters should be considered, including resources and

limitations. The business objective should be translated into a data mining goal. Success criteria should be defined and a project plan should be developed.

2. **Data understanding:** This phase involves considering the data requirements for properly addressing the defined goal and an investigation of the availability of the required data. This phase also includes initial data collection and exploration with summary statistics and visualization tools to understand the data and identify potential problems in availability and quality.

3. **Data preparation:** The data to be used should be identified, selected, and prepared for inclusion in the data mining model. This phase involves the acquisition, integration, and formatting of the data according to the needs of the project. The consolidated data should then be "cleaned" and properly transformed according to the requirements of the algorithm to be applied. New fields such as sums, averages, ratios, flags, and so on should be derived from the raw fields to enrich customer information, to better summarize customer characteristics, and therefore to enhance the performance of the models.

4. **Modeling:** The processed data are then used for model training. Analysts should select the appropriate modeling technique for the particular business objective. Before the training of the models and especially in the case of predictive modeling, the modeling dataset should be partitioned so that the model's performance is evaluated on a separate dataset. This phase involves the examination of alternative modeling algorithms and parameter settings and a comparison of their fit and performance in order to find the one that yields the best results. Based on an initial evaluation of the model results, the model settings can be revised and fine tuned.

5. **Evaluation:** The generated models are then formally evaluated not only in terms of technical measures but also, more importantly, in the context of the business success criteria set out in the business understanding phase. The project team should decide whether the results of a given model properly address the initial business objectives. If so, this model is approved and prepared for deployment.

6. **Deployment:** The project's findings and conclusions are summarized in a report, but this is hardly the end of the project. Even the best model will turn out to be a business failure if its results are not deployed and integrated into the organization's everyday marketing operations. A procedure should be designed and developed to enable the scoring of customers and the updating of the results. The deployment procedure should also enable the distribution of the model results throughout the enterprise and their incorporation in the organization's databases and operational CRM system. Finally, a maintenance plan should be designed and the whole process should be reviewed. Lessons learned should be taken into account and the next steps should be planned.

The phases above present strong dependencies and the outcomes of a phase may lead to revisiting and reviewing the results of preceding phases. The nature of the process is cyclical since the data mining itself is a never-ending journey and quest, demanding continuous reassessment and updating of completed tasks in the context of a rapidly changing business environment.

DATA MINING AND BUSINESS DOMAIN EXPERTISE

The role of data mining models in marketing is quite new. Although rapidly expanding, data mining is still "foreign territory" for many marketers who trust only their "intuition" and domain experience. Their segmentation schemes and marketing campaign lists are created by business rules based on their business knowledge.

Data mining models are not "threatening": they cannot substitute or replace the significant role of domain experts and their business knowledge. These models, however powerful, cannot effectively work without the active support of business experts. On the contrary, only when data mining capabilities are complemented with business expertise can they achieve truly meaningful results. For instance, the predictive ability of a data mining model can be substantially increased by including informative inputs with predictive power suggested by persons with experience in the field. Additionally, the information of existing business rules/scores can be integrated into a data mining model and contribute to the building of a more robust and successful result. Moreover, before the actual deployment, model results should always be evaluated by business experts with respect to their meaning, in order to minimize the risk of coming up with trivial or unclear findings. Thus, business domain knowledge can truly help and enrich the data mining results.

On the other hand, data mining models can identify patterns that even the most experienced business people may have missed. They can help in fine tuning the existing business rules, and enrich, automate, and standardize judgmental ways of working which are based on personal perceptions and views. They comprise an objective, data-driven approach, minimizing subjective decisions and simplifying time-consuming processes.

In conclusion, the combination of business domain expertise with the power of data mining models can help organizations gain a competitive advantage in their efforts to optimize customer management.

SUMMARY

In this chapter we introduced data mining. We presented the main types of data mining models and a process model, a methodological framework for designing and implementing successful data mining projects. We also outlined how data mining can help an organization to better address the CRM objectives and achieve "individualized" and more effective customer management through customer insight. The following list summarizes some of the most useful data mining applications in the CRM framework:

- Customer segmentation:

  - **Value-based segmentation:** Customer ranking and segmentation according to current and expected/estimated customer value.
  - **Behavioral segmentation:** Customer segmentation based on behavioral attributes.
  - **Value-at-risk segmentation:** Customer segmentation based on value and estimated voluntary churn propensity scores.

- Targeted marketing campaigns:

  - Voluntary churn modeling and estimation of the customer's likelihood/ propensity to churn.
  - Estimation of the likelihood/propensity to take up an add-on product, to switch to a more profitable product, or to increase usage of an existing product.
  - Estimation of the lifetime value (LTV) of customers.

Table 1.2 presents some of the most widely used data mining modeling techniques together with an indicative listing of the marketing applications they can support.

Table 1.2   Data mining modeling techniques and their applications.

Category of modeling techniques	Modeling techniques	Applications
Classification (propensity) models	Neural networks, decision trees, logistic regression, etc.	• Voluntary churn prediction • Cross/up/deep selling
Clustering models	K-means, TwoStep, Kohonen network/self-organizing map, etc.	• Segmentation
Association and sequence models	A priori, Generalized Rule Induction, sequence	• Market basket analysis • Web path analysis

The next two chapters are dedicated to data mining modeling techniques. The first one provides a brief introduction to the main modeling concepts and aims to familiarize the reader with the most widely used techniques. The second one goes a step further and focuses on the techniques used for segmentation. As the main scope of the book is customer segmentation, these techniques are presented in detail and step by step, preparing readers for executing similar projects on their own.

CHAPTER TWO

An Overview of Data Mining Techniques

SUPERVISED MODELING

In supervised modeling, whether for the prediction of an event or for a continuous numeric outcome, the availability of a training dataset with historical data is required. Models learn from past cases. In order for predictive models to associate input data patterns with specific outcomes, it is necessary to present them with cases with known outcomes. This phase is called the training phase. During that phase, the predictive algorithm builds the function that connects the inputs with the target field. Once the relationships are identified and the model is evaluated and proved to be of satisfactory predictive power, the scoring phase follows. New records, for which the outcome values are unknown, are presented to the model and scored accordingly.

Some predictive models such as regression and decision trees are transparent, providing an explanation of their results. Besides prediction, these models can also be used for insight and profiling. They can identify inputs with a significant effect on the target attribute and they can reveal the type and magnitude of the effect. For instance, supervised models can be applied to find the drivers associated with customer satisfaction or attrition. Similarly, supervised models can also supplement traditional reporting techniques in the profiling of the segments of an organization by identifying the differentiating features of each group.

Data Mining Techniques in CRM: Inside Customer Segmentation K. Tsiptsis and A. Chorianopoulos
© 2009 John Wiley & Sons, Ltd

According to the measurement level of the field to be predicted, supervised models are further categorized into:

- **Classification or propensity modeling techniques.**
- **Estimation or regression modeling techniques.**

A categorical or symbolic field contains discrete values which denote membership of known groups or predefined classes. A categorical field may be a flag (dichotomous or binary) field with Yes/No or True/False values or a set field with more than two outcomes. Typical examples of categorical fields and outcomes include:

- Accepted a marketing offer. [Yes/No]
- Good credit risk/bad credit risk.
- Churned/stayed active.

These outcomes are associated with the occurrence of specific events. When the target is categorical, the use of a classification model is appropriate. These models analyze discrete outcomes and are used to classify new records into the predefined classes. In other words, they predict events. Confidence scores supplement their predictions, denoting the likelihood of a particular outcome.

On the other hand, there are fields with continuous numeric values (range values), such as:

- The balance of bank accounts
- The amount of credit card purchases of each card holder
- The number of total telecommunication calls made by each customer.

In such cases, when analysts want to estimate continuous outcome values, estimation models are applied. These models are also referred to as regression models after the respective statistical technique. Nowadays, though, other estimation techniques are also available.

Another use of supervised models is in the screening of predictors. These models are used as a preparatory step before the development of a predictive model. They assess the predictive importance of the original input fields and identify the significant predictors. Predictors with little or no predictive power are removed from the subsequent modeling steps.

The different uses of supervised modeling techniques are depicted in Figure 2.1.

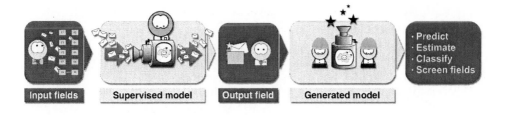

Figure 2.1   Graphical representation of supervised modeling.

PREDICTING EVENTS WITH CLASSIFICATION MODELING

As described above, classification models predict categorical outcomes by using a set of input fields and a historical dataset with pre-classified data. Generated models are then used to predict the occurrence of events and classify unseen records. The general idea of a classification models is described in the next, simplified example.

A mobile telephony network operator wants to conduct an outbound cross-selling campaign to promote an Internet service to its customers. In order to optimize the campaign results, the organization is going to offer the incentive of a reduced service cost for the first months of usage. Instead of addressing the offer to the entire customer base, the company decided to target only prospects with an increased likelihood of acceptance. Therefore it used data mining in order to reveal the matching customer profile and identify the right prospects. The company decided to run a test campaign in a random sample of its existing customers which currently were not using the Internet service. The campaign's recorded results define the output field. The input fields include all the customer demographics and usage attributes which already reside in the organization's data mart.

Input and output fields are joined into a single dataset for the purposes of model building. The final form of the modeling dataset, for eight imaginary customers and an indicative list of inputs (gender, occupation category, volume/traffic of voice and SMS usage), is shown in Table 2.1.

The classification procedure is depicted in Figure 2.2.

The data are then mined with a classification model. Specific customer profiles are associated with acceptance of the offer. In this simple, illustrative example, none of the two contacted women accepted the offer. On the other hand, two out of the five contacted men (40%) were positive toward the offer. Among white-collar men this percentage reaches 67% (two out of three). Additionally, all white-collar men with heavy SMS usage turned out to be interested in the Internet service. These customers comprise the service's target group. Although oversimplified, the described process shows the way that classification algorithms work. They analyze predictor fields and map input data patterns with specific outcomes.

Table 2.1 The modeling dataset for the classification model.

Customer ID	Gender	Occupation	Monthly average number of SMS calls	Monthly average number of voice calls	Response to pilot campaign
		Input fields			Output field
1	Male	White collar	28	140	No
2	Male	Blue collar	32	54	No
3	Female	Blue collar	57	30	No
4	Male	White collar	143	140	Yes
5	Female	White collar	87	81	No
6	Male	Blue collar	143	28	No
7	Female	White collar	150	140	No
8	Male	White collar	140	60	Yes

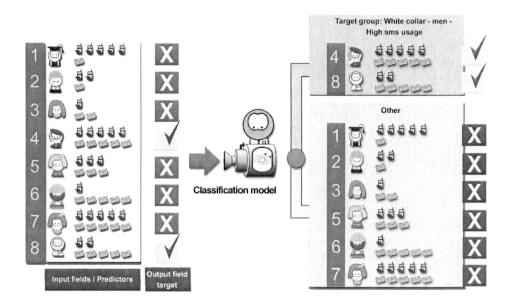

Figure 2.2 Graphical representation of classification modeling.

After identifying the customer profiles associated with acceptance of the offer, the company extrapolated the results to the whole customer base to construct a campaign list of prospective Internet users. In other words, it scored all customers with the derived model and classified customers as potential buyers or non-buyers.

In this naive example, the identification of potential buyers could also be done with inspection by eye. But imagine a situation with hundreds of candidate predictors and tens of thousands of records or customers. Such complicated but realistic tasks which human brains cannot handle can be easily and effectively carried out by data mining algorithms.

What If There Is Not an Explicit Target Field to Predict?

In some cases there is no apparent categorical target field to predict. For example, in the case of prepaid customers in mobile telephony, there is no recorded disconnection event to be modeled. The separation between active and churned customers is not evident. In such cases a target event could be defined with respect to specific customer behavior. This handling requires careful data exploration and co-operation between the data miners and the marketers. For instance, prepaid customers with no incoming or outgoing phone usage within a certain time period could be considered as churners. In a similar manner, certain behaviors or changes in behavior, for instance a substantial decrease in usage or a long period of inactivity, could be identified as signals of specific events and then used for the definition of the respective target. Moreover, the same approach could also be followed when analysts want to act proactively. For instance, even when a churn/disconnection event could be directly identified through a customer's action, a proactive approach would analyze and model customers before their typical attrition, trying to identify any early signals of defection and not waiting for official termination of the relationship with the customer.

At the heart of all classification models is the estimation of confidence scores. These are scores that denote the likelihood of the predicted outcome. They are estimates of the probability of occurrence of the respective event. The predictions generated by the classification models are based on these scores: a record is classified into the class with the largest estimated confidence. The scores are expressed on a continuous numeric scale and usually range from 0 to 1. Confidence scores are typically translated to propensity scores which signify the likelihood of a particular outcome: the propensity of a customer to churn, to buy a specific add-on product, or to default on a loan. Propensity scores allow for the rank ordering of customers according to the likelihood of an outcome. This feature enables marketers to tailor the size of their campaigns according to their resources and marketing objectives. They can expand or reduce their target lists on the basis of their particular objectives, always targeting those customers with the relatively higher probabilities.

The purpose of all classification models is to provide insight and help in the refinement and optimization of marketing applications. The first step after model training is to browse the generated results, which may come in different forms according to the model used: rules, equations, graphs. Knowledge extraction is followed by evaluation of the model's predictive efficiency and by the deployment of the results in order to classify new records according to the model's findings. The whole procedure is described in Figure 2.3, which is explained further below.

The following modeling techniques are included in the class of classification models:

- **Decision trees:** Decision trees operate by recursively splitting the initial population. For each split they automatically select the most significant predictor, the predictor that yields the best separation with respect to the target field. Through successive partitions, their goal is to produce "pure" sub-segments, with homogeneous behavior in terms of the output. They are perhaps the most popular classification technique. Part of their popularity is because they produce transparent results that are easily interpretable, offering an insight into the event under study. The produced results can have two equivalent formats. In a rule format, results are represented in plain English as ordinary rules:

```
IF (PREDICTOR VALUES) THEN (TARGET OUTCOME AND CONFIDENCE SCORE).
```

  For example:

```
IF (Gender=Male and Profession=White Collar and SMS_Usage > 60
messages per month) THEN Prediction=Buyer and Confidence=0.95.
```

  In a tree format, rules are graphically represented as a tree in which the initial population (root node) is successively partitioned into terminal nodes or leaves of sub-segments with similar behavior in regard to the target field.

  Decision tree algorithms provide speed and scalability. Available algorithms include:

  - C5.0
  - CHAID
  - Classification and Regression Trees
  - QUEST.

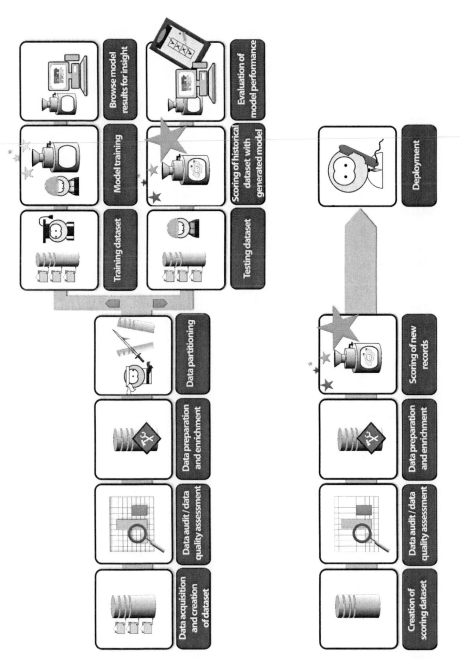

Figure 2.3   An outline of the classification modeling procedure.

- **Decision rules:** These are quite similar to decision trees and produce a list of rules which have the format of human-understandable statements:

```
IF (PREDICTOR VALUES) THEN (TARGET OUTCOME AND CONFIDENCE SCORE).
```

Their main difference from decision trees is that they may produce multiple rules for each record. Decision trees generate exhaustive and mutually exclusive rules which cover all records. For each record only one rule applies. On the contrary, decision rules may generate an overlapping set of rules. More than one rule, with different predictions, may hold true for each record. In that case, rules are evaluated, through an integrated procedure, to determine the one for scoring. Usually a voting procedure is applied, which combines the individual rules and averages their confidences for each output category. Finally, the category with the highest average confidence is selected as the prediction. Decision rule algorithms include:

- C5.0
- Decision list.

- **Logistic regression:** This is a powerful and well-established statistical technique that estimates the probabilities of the target categories. It is analogous to simple linear regression but for categorical outcomes. It uses the generalized linear model and calculates regression coefficients that represent the effect of predictors on the probabilities of the categories of the target field. Logistic regression results are in the form of continuous functions that estimate the probability of membership in each target outcome:

$$\ln(p/(1-p)) = b_0 + b_1 \cdot \text{Predictor } 1 + b_2 \cdot \text{Predictor } 2 \\ + \cdots + b_n \cdot \text{Predictor } N$$

where $p = $ probability of an event to happen.
For example:

$$\ln(\text{churn probability}/(\text{no churn probability})) \\ = b_0 + b_1 \cdot \text{Tenure} + b_2 \cdot \text{Number of products} + \cdots.$$

In order to yield optimal results it may require special data preparation, including potential screening and transformation of the predictors. It still demands some statistical experience, but provided it is built properly it can produce stable and understandable results.

- **Neural networks:** Neural networks are powerful machine learning algorithms that use complex, nonlinear mapping functions for estimation and classification.

They consist of neurons organized in layers. The input layer contains the predictors or input neurons. The output layer includes the target field. These models estimate weights that connect predictors (input layer) to the output. Models with more complex topologies may also include intermediate, hidden layers, and neurons. The training procedure is an iterative process. Input records, with known outcomes, are presented to the network and model prediction is evaluated with respect to the observed results. Observed errors are used to adjust and optimize the initial weight estimates. They are considered as opaque or "black box" solutions since they do not provide an explanation of their predictions. They only provide a sensitivity analysis, which summarizes the predictive importance of the input fields. They require minimum statistical knowledge but, depending on the problem, may require a long processing time for training.

- **Support vector machine(SVM):** SVM is a classification algorithm that can model highly nonlinear, complex data patterns and avoid overfitting, that is, the situation in which a model memorizes patterns only relevant to the specific cases analyzed. SVM works by mapping data to a high-dimensional feature space in which records become more easily separable (i.e., separated by linear functions) with respect to the target categories. Input training data are appropriately transformed through nonlinear kernel functions and this transformation is followed by a search for simpler functions, that is, linear functions, which optimally separate records. Analysts typically experiment with different transformation functions and compare the results. Overall SVM is an effective yet demanding algorithm, in terms of memory resources and processing time. Additionally, it lacks transparency since the predictions are not explained and only the importance of predictors is summarized.

- **Bayesian networks:** Bayesian models are probability models that can be used in classification problems to estimate the likelihood of occurrences. They are graphical models that provide a visual representation of the attribute relationships, ensuring transparency, and an explanation of the model's rationale.

Evaluation of Classification Models

Before applying the generated model in new records, an evaluation procedure is required to assess its predictive ability. The historical data with known outcomes, which were used for training the model, are scored and two new fields are derived: the predicted outcome category and the respective confidence score, as shown in Table 2.2, which illustrates the procedure for the simplified example presented earlier.

In practice, models are never as accurate as in the simple exercise presented here. There are always errors and misclassified records. A comparison of the predicted to the actual values is the first step in evaluating the model's performance.

Table 2.2 Historical data and model-generated prediction fields.

			Input fields		Output field	Model-generated fields	
Customer ID	Gender	Profession	Monthly average number of SMS calls	Monthly average number of voice calls	Response to pilot campaign	Predicted response	Estimated response confidence score
1	Male	White collar	28	140	No	No	0.0
2	Male	Blue collar	32	54	No	No	0.0
3	Female	Blue collar	57	30	No	No	0.0
4	Male	White collar	143	140	Yes	Yes	1.0
5	Female	White collar	87	81	No	No	0.0
6	Male	Blue collar	143	28	No	No	0.0
7	Female	White collar	150	140	No	No	0.0
8	Male	White collar	140	60	Yes	Yes	1.0

This comparison provides an estimate of the model's future predictive accuracy on unseen cases. In order to make this procedure more valid, it is advisable to evaluate the model in a dataset that was not used for training the model. This is achieved by partitioning the historical dataset into two distinct parts through random sampling: the training and the testing dataset. A common practice is to allocate approximately 70–75% of the cases to the training dataset. Evaluation procedures are applied to both datasets. Analysts should focus mainly on the examination of performance indicators in the testing dataset. A model underperforming in the testing dataset should be re-examined since this is a typical sign of overfitting and of memorizing the specific training data. Models with this behavior do not provide generalizable results. They provide solutions that only work for the particular data on which they were trained.

Some analysts use the testing dataset to refine the model parameters and leave a third part of the data, namely the validation dataset, for evaluation. However, the best approach, which unfortunately is not always employed, would be to test the model's performance in a third, disjoint dataset from a different time period.

One of the most common performance indicators for classification models is the error rate. It measures the percentage of misclassifications. The overall error rate indicates the percentage of records that were not correctly classified by the model. Since some mistakes may be more costly than others, this percentage is also estimated for each category of the target field. The error rate is summarized in misclassification or coincidence or confusion matrices that have the form given in Table 2.3.

Table 2.3   Misclassification matrix.

		Predicted values	
		Positive	**Negative**
Actual values	**Positive**	Correct prediction: true positive record count	Misclassification: false negative record count
	Negative	Misclassification: false positive record count	Correct prediction: true negative record count

The gains, response, and lift/index tables and charts are also helpful evaluation tools that can summarize the predictive efficiency of a model with respect to a specific target category. To illustrate their basic concepts and usage we will present the results of a hypothetical churn model that was built on a dichotomous output field which flagged churners.

The first step in the creation of such charts and tables is to select the target category of interest, also referred to as the hit category. Records/customers are then ordered according to their hit propensities and binned into groups of equal size, named quantiles. In our hypothetical example, the target is the category of churners and the hit propensity is the churn propensity; in other words, the estimated likelihood of belonging to the group of churners. Customers have been split into 10 equal groups of 10% each, named deciles. The 10% of customers with the highest churn propensities comprise tile 1 and those with the lowest churn propensities, tile 10. In general, we expect that high estimated hit propensities also correspond to the actual customers of the target category. Therefore, we hope to find large concentrations of actual churners among the top model tiles.

The cumulative table, Table 2.4, evaluates our churn model in terms of the gain, response, and lift measures.

But what exactly do these performance measures represent and how are they used for model evaluation? A brief explanation is as follows:

- **Response %:** "How likely is the target category within the examined quantiles?" Response % denotes the percentage (probability) of the target category within the quantiles. In our example, 10.7% of the customers of the top 10% model tile were actual churners, yielding a response % of the same value. Since the overall churn rate was 2.9%, we expect that a random list would also have an analogous churn rate. However, the estimated churn rate for the top model tile was 3.71 times (or 371.4%) higher. This is called the lift. Analysts have achieved results about four times better than randomness in the examined model tile. As we move from the top to the bottom tiles, the model estimated confidences decrease.

Table 2.4 The gains, response, and lift table.

Model tiles	Cumulative % of records	Gain %	Response %	Lift (%)
1	10	37.1	10.7	371.4
2	20	56.9	8.2	284.5
3	30	69.6	6.7	232.1
4	40	79.6	5.7	199.0
5	50	87.0	5.0	174.1
6	60	91.6	4.4	152.7
7	70	94.6	3.9	135.2
8	80	96.4	3.5	120.6
9	90	98.2	3.1	109.2
10	100	100.0	2.9	100.0

The concentration of the actual churners is also expected to decrease. Indeed, the first two tiles, which jointly account for the top 20% of customers with the highest estimated churn scores, have a smaller percentage of actual churners (8.2%). This percentage is still 2.8 times higher than randomness, though.

- **Gain %:** "How many of the target population fall in the quantiles?" Gain % is defined as the percentage of the total target population that belongs in the quantiles. In our example, the top 10% model tile contains 37.1% of all actual churners, yielding a gain % of the same value. A random list containing 10% of the customers would normally capture about 10% of all observed churners. However, the top model tile contains more than a third (37.1%) of all observed churners. Once again we come to the lift concept. The top 10% model tile identifies about four times more target customers than a random list of the same size.

- **Lift:** "How much better are the model results compared to randomness?" The lift or index assesses the improvement in predictive ability due to the model. It is defined as the ratio of the response % to the prior probability. In other words, it compares the model quantiles to a random list of the same size in terms of the probability of the target category. Therefore it represents how much a data mining model exceeds the baseline model of random selection.

The gain, response, and lift evaluation measures can also be depicted in corresponding charts such as those shown below. The two added reference lines correspond to the top 5% and the top 10% tiles. The diagonal line in the gains chart represents the baseline model of randomness.

The response chart (Figure 2.4) visually illustrates the estimated churn probability among the mode tiles. As we move to the left of the X-axis and toward the top tiles, we have increased churn probabilities. These tiles would result in more

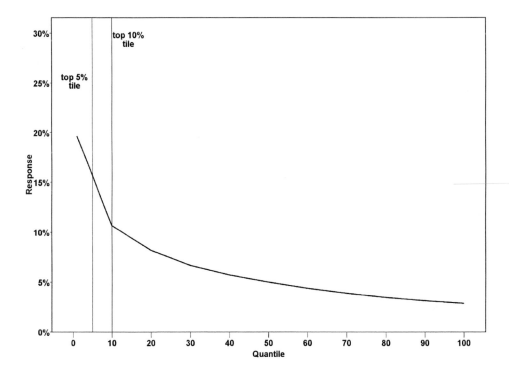

Figure 2.4   Response chart.

targeted lists and smaller error rates. Expanding the list to the right of the X-axis, toward the bottom model tiles, would increase the expected false positive error rate by including in the targeting list more customers with no real intention to churn.

According to the gains chart (Figure 2.5), when scoring an unseen customer list, data miners should expect to capture about 40% of all potential churners if they target the customers of the top 10% model tile. Narrowing the list to the top 5% tile decreases the percentage of potential churners to be reached to approximately 25%. As we move to the right of the X-axis, the expected number of total churners to be identified increases. At the same time, though, as we have seen in the response chart, the respective error rate of false positives increases. On the contrary, the left parts of the X-axis lead to smaller but more targeted campaigns.

The lift or index chart (Figure 2.6) directly compares the model's predictive performance to the baseline model of random selection. The concentration of churners is estimated to be four times higher than randomness among the top 10% customers and about six times higher among the top 5% customers.

By studying these charts marketers can gain valuable insight into the model's future predictive accuracy on new records. They can then decide on the size of the

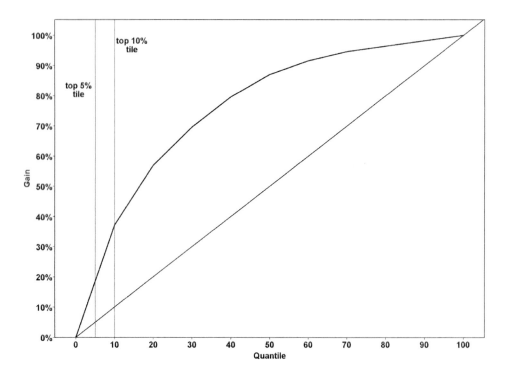

Figure 2.5   Gains chart.

respective campaign by choosing the tiles to target. They may choose to conduct a small campaign, limited to the top tiles, in order to address only those customers with very high propensities and minimize the false positive cases. Alternatively, especially if the cost of the campaign is small compared to the potential benefits, they may choose to expand their list by including more tiles and more customers with relatively lower propensities.

In conclusion, these charts can answer questions such as:

- What response rates should we expect if we target the top $n\%$ of customers according to the model-estimated propensities?
- How many target customers (potential churners or buyers) are we about to identify by building a campaign list based on the top $n\%$ of the leads according to the model?

The answers permit marketers to build scenarios on different campaign sizes. The estimated results may include more information than just the expected response rates. Marketers can incorporate cost and revenue information and build profit and ROI (Return On Investment) charts to assess their upcoming campaigns in terms of expected cost and revenue.

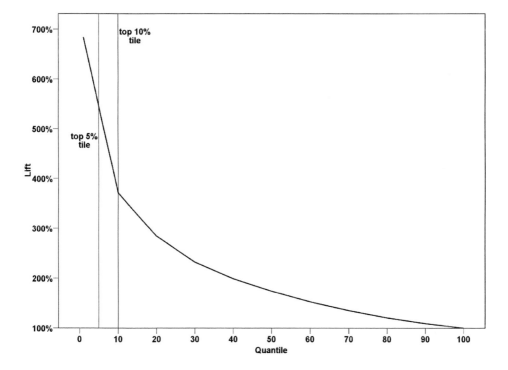

Figure 2.6   Lift chart.

The Maximum Benefit Point

An approach often referred to in the literature as a rule of thumb for selecting the optimal size of a targeted marketing campaign list is to examine the gains chart and select all top tiles up to the point where the distance between the gains curve and the diagonal reference line becomes a maximum. This is referred to as the maximum benefit point and it is the point where the difference between the gains curve and the diagonal reference line has its maximum value. The reasoning behind this approach is that, from that point on, the model classifies worse than randomness. This approach usually yields large targeting lists. In practice analysts and marketers should take into consideration the particular business situation, objectives, and resources and possibly consider as a classification threshold the point of lift maximization. If possible, they should also incorporate in the gains chart cost (per offer) and revenue (per acceptance) information and select the cut-point that best serves their specific business needs and maximizes the expected ROI and profit.

Scoring with Classification Models

Once the classification model is trained and evaluated, the next step is to deploy it and use the generated results to develop and carry out direct marketing campaigns. Each model, apart from offering insight through the revealed data patterns, can also be used as a scoring engine. When unseen data are passed through the derived model, they are scored and classified according to their estimated confidence scores.

As we saw above, the procedure for assigning records to the predefined classes may not be left entirely to the model specifications. Analysts can consult the gains charts and intervene in the predictions by setting a classification threshold that best serves their needs and their business objectives. Thus, they can expand or decrease the size of the derived marketing campaign lists according to the expected response rates and the requirements of the specific campaign.

The actual response rates of the executed campaigns should be monitored and evaluated. The results should be recorded in campaign libraries as they could be used for training relevant models in the future.

Finally, an automated and standardized procedure should be established that will enable the updating of the scores and their loading into the existing campaign management systems.

MARKETING APPLICATIONS SUPPORTED BY CLASSIFICATION MODELING

Marketing applications aim at establishing a long-term and profitable relationship with customers, throughout the whole lifetime of the customer. Classification models can play a significant role in marketing, specifically in the development of targeted marketing campaigns for acquisition, cross/up/deep selling, and retention. Table 2.5 presents a list of these applications along with their business objectives.

All the above applications can be supported by classification modeling. A classification model can be applied to identify the target population and recognize customers with an increased likelihood for churn or additional purchase. In other words, the event of interest (acquisition, churn, cross/up/deep selling) can be translated into a categorical target field which can then be used as an output in a classification model. Targeted campaigns can then be conducted with contact lists based on data mining models.

Setting up a data mining procedure for the needs of these applications requires special attention and co-operation between data miners and marketers.

Table 2.5 Marketing application and campaigns that can be supported by classification modeling.

Business objective	Marketing application
Getting customers	• Acquisition: finding new customers and expanding the customer base with new and potentially profitable customers
Developing customers	• Cross selling: promoting and selling additional products or services to existing customers • Up selling: offering and switching customers to premium products, other products more profitable than the ones that they already have • Deep selling: increasing usage of the products or services that customers already have
Retaining customers	• Retention: prevention of voluntary churn, with priority given to presently or potentially valuable customers

The most difficult task is usually to decide on the target event and population. The analysts involved should come up with a valid definition that makes business sense and can lead to really effective and proactive marketing actions. For instance, before starting to develop a churn model we should have an answer to the "what constitutes churn?" question. Even if we build a perfect model, this may turn out to be a business failure if, due to our target definition, it only identifies customers who are already gone by the time the retention campaign takes place.

Predictive modeling and its respective marketing applications are beyond the scope of this book, which focuses on customer segmentation. Thus, we will not deal with these important methodological issues here. In the next section, though, we will briefly outline an indicative methodological approach for setting up a voluntary churn model.

SETTING UP A VOLUNTARY CHURN MODEL

In this simplified example, the goal of a mobile telephony network operator is to set up a model for the early identification of potential voluntary churners. This model will be the base for a respective targeted retention campaign and predicts voluntary attrition three months ahead. Figure 2.7 presents the setup.

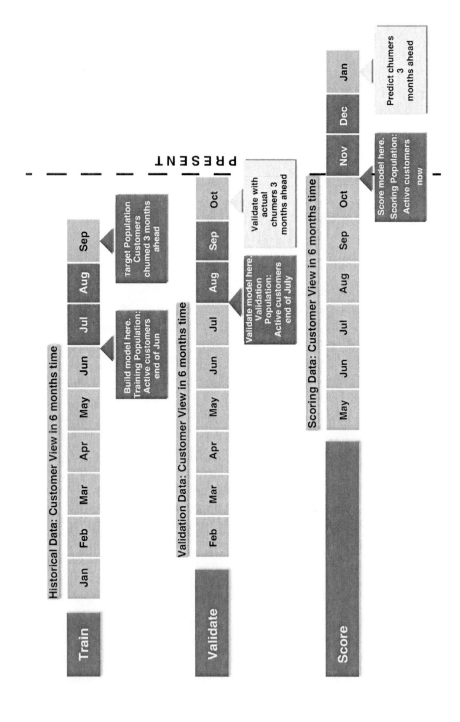

Figure 2.7   Setting up a voluntary churn model.

The model is trained on a six-month historical dataset. The methodological approach is outlined by the following points:

- The input fields used cover all aspects of the customer relationship with the organization: customer and contract characteristics, usage and behavioral indicators, and so on, providing an integrated customer view also referred to as the customer signature.
- The model is trained on customers who were active at the end of the historical period (end of the six-month period). These customers comprise the training population.
- A three-months period is used for the definition of the target event and the target population.
- The target population consists of those who have voluntary churned (applied for disconnection) by the end of the three-month period.
- The model is trained by identifying the input data patterns (customer characteristics) associated with voluntary churn.
- The generated model is validated on a disjoint dataset of a different time period, before being deployed for scoring presently active customers.
- In the deployment or scoring phase, presently active customers are scored according to the model and churn propensities are generated. The model predicts churn three months ahead.
- The generated churn propensities can then be used for better targeting of an outbound retention campaign. The churn model results can be combined and cross-examined with the present or potential value of the customers so that the retention activities are prioritized accordingly.
- All input data fields that were used for the model training are required, obviously with refreshed information, in order to update the churn propensities.
- Two months have been reserved to allow for scoring and preparing the campaign. These two months are shown as gray boxes in the figure and are usually referred to as the latency period.
- A latency period also ensures that the model is not trained to identify "immediate" churners. Even if we manage to identify those customers, the chances are that by the time they are contacted, they could already be gone or it will be too late to change their minds. The goal of the model should be long term: the recognition of early churn signals and the identification of customers with an increased likelihood to churn in the near but not immediate future, since for them there is a chance of retention.
- To build a long-term churn model, immediate churners, namely customers who churned during the two-month latency period, are excluded from the model training.

- The definition of the target event and the time periods used in this example are purely indicative. A different time frame for the historical or latency period could be used according to the specific task and business situation.

FINDING USEFUL PREDICTORS WITH SUPERVISED FIELD SCREENING MODELS

Another class of supervised modeling techniques includes the supervised field screening models (Figure 2.8). These are models that usually serve as a preparation step for the development of classification and estimation models. The situation of having hundreds or even thousands of candidate predictors is not an unusual one in complicated data mining tasks. Some of these fields, though, may not have an influence on the output field that we want to predict. The role of supervised field screening models is to assess all the available inputs and find the key predictors and those predictors with marginal or no importance that are candidates for potential removal from the predictive model.

   Some predictive algorithms, including decision trees, for example, integrate screening mechanisms that internally filter out the unrelated predictors. There are some other algorithms which are inefficient when handling a large number of candidate predictors at reasonable times. The field screening models can efficiently reduce data dimensionality, retaining only those fields relevant to the

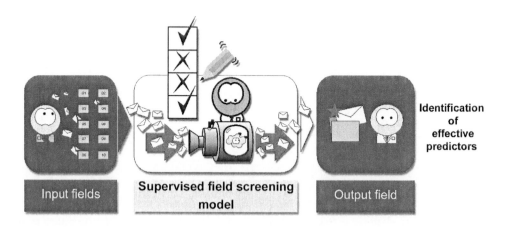

Figure 2.8   Supervised field screening models.

outcome of interest, allowing data miners to focus only on the information that matters.

Field screening models are usually used in the data preparation phase of a data mining project in order to perform the following tasks:

- Evaluate the quality of potential predictors. They incorporate specific criteria to identify inadequate predictors: for instance, predictors with an extensive percentage of missing (null) values, continuous predictors which are constant or have little variation, categorical predictors with too many categories or with almost all records falling in a single category.
- Rank predictors according to their predictive power. The influence of each predictor on the target field is assessed and an importance measure is calculated. Predictors are then sorted accordingly.
- Filter out unimportant predictors. Predictors unrelated to the target field are identified. Analysts have the option to filter them out, reducing the set of input fields to those related to the target field.

PREDICTING CONTINUOUS OUTCOMES WITH ESTIMATION MODELING

Estimation models, also referred to as regression models, deal with continuous numeric outcomes. By using linear or nonlinear functions they use the input fields to estimate the unknown values of a continuous target field.

Estimation techniques can be used to predict attributes like the following:

- The expected balance of the savings accounts of bank customers in the near future.
- The estimated volume of traffic for new customers of a mobile telephony network operator.
- The expected revenue from a customer for the next year.

A dataset with historical data and known values of the continuous output is required for training the model. A mapping function is then identified that associates the available inputs to the output values. These models are also referred to as regression models, after the well-known and established statistical technique of ordinary least squares regression (OLSR), which estimates the line that best fits the data and minimizes the observed errors, the so-called least squares line. It requires some statistical experience and since it is sensitive to possible violations of its assumptions it may require specific data examination and processing before building. The final model has the intuitive form of a linear function with coefficients denoting the effect of predictors on the outcome measure. Although transparent, it has inherent limitations that may affect its predictive performance in complex situations of nonlinear relationships and interactions between predictors.

Nowadays, traditional regression is not the only available estimation technique. New techniques, with less stringent assumptions and which also capture nonlinear relationships, can also be employed to handle continuous outcomes. More specifically, neural networks, SVM, and specific types of decision trees, such as Classification and Regression Trees and CHAID, can also be employed for the prediction of continuous measures.

The data setup and the implementation procedure of an estimation model are analogous to those of a classification model. The historical dataset is used for training the model. The model is evaluated with respect to its predictive effectiveness, in a disjoint dataset, preferably of a different time period, with known outcome values. The generated model is then deployed on unseen data to estimate the unknown target values.

The model creates one new field when scoring: the estimated outcome value. Estimation models are evaluated with respect to the observed errors: the deviation, the difference between the predicted and the actual values. Errors are also called residuals.

A large number of residual diagnostic plots and measures are usually examined to assess the model's predictive accuracy. Error measures typically examined include:

- Correlation measures between the actual and the predicted values, such as the Pearson correlation coefficient. This coefficient is a measure of the linear association between the observed and the predicted values. Values close to 1 indicate a strong relationship and a high degree of association between what was predicted and what is really happening.
- The relative error. This measure denotes the ratio of the variance of the observed values from those predicted by the model to the variance of the observed values from their mean. It compares the model with a baseline model that simply returns the mean value as the prediction for all records. Small values indicate better models. Values greater than 1 indicate models less accurate than the baseline model and therefore not useful.
- Mean error or mean squared error across all examined records.
- Mean absolute error (MAE).
- Mean absolute percent error (MAPE).

Examining the Model Errors to Reveal Anomalous or Even Suspect Cases

The examination of deviations of the predicted from the actual values can also be used to identify outlier or abnormal cases. These cases may simply indicate poor model performance or an unusual but acceptable behavior. Nevertheless, they deserve special inspection since they may also be signs of suspect behavior.

For instance, an insurance company can build an estimation model based on the amounts of claims by using the claim application data as predictors. The resulting model can then be used as a tool to detect fraud. Entries that substantially deviate from the expected values could be identified and further examined or even sent to auditors for manual inspection.

UNSUPERVISED MODELING TECHNIQUES

In the previous sections we briefly presented the supervised modeling techniques. Whether used for classification, estimation, or field screening, their common characteristic is that they all involve a target attribute which must be associated with an examined set of inputs. The model training and data pattern recognition are guided or supervised by a target field. This is not the case in unsupervised modeling, in which only input fields are involved. All inputs are treated equally in order to extract information that can be used, mainly, for the identification of groupings and associations.

Clustering techniques identify meaningful natural groupings of records and group customers into distinct segments with internal cohesion. Data reduction techniques like factor analysis or principal components analysis (PCA) "group" fields into new compound measures and reduce the data's dimensionality without losing much of the original information. But grouping is not the only application of unsupervised modeling. Association or affinity modeling is used to discover co-occurring events, such as purchases of related products. It has been developed as a tool for analyzing shopping cart patterns and that is why it is also referred to as market basket analysis. By adding the time factor to association modeling we have sequence modeling: in sequence modeling we analyze associations over time and try to discover the series of events, the order in which events happen. And that is not all. Sometimes we are just interested in identifying records that "do not fit well," that is, records with unusual and unexpected data patterns. In such cases, record screening techniques can be employed as a data auditing step before building a subsequent model to detect abnormal (anomalous) records.

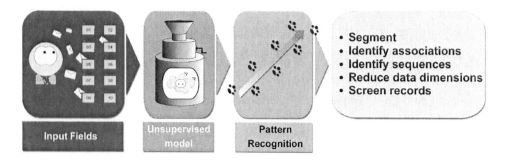

Figure 2.9   Graphical representation of unsupervised modeling.

Below, we will briefly present all these techniques before focusing on the clustering and data reduction techniques used mainly for segmentation purposes.

The different uses of supervised modeling techniques are depicted in Figure 2.9.

SEGMENTING CUSTOMERS WITH CLUSTERING TECHNIQUES

Consider the situation of a social gathering where guests start to arrive and mingle with each other. After a while, guests start to mix in company and groups of socializing people start to appear. These groups are formed according to the similarities of their members. People walk around and join groups according to specific criteria such as physical appearance, dress code, topic and tone of discussion, or past acquaintance. Although the host of the event may have had some initial presumptions about who would match with whom, chances are that at the end of the night some quite unexpected groupings would come up.

Grouping according to proximity or similarity is the key concept of clustering. Clustering techniques reveal natural groupings of "similar" records. In the small stores of old, when shop owners knew their customers by name, they could handle all clients on an individual basis according to their preferences and purchase habits. Nowadays, with thousands or even millions of customers, this is not feasible. What is feasible, though, is to uncover the different customer types and identify their distinct profiles. This constitutes a large step on the road from mass marketing to a more individualized handling of customers. Customers are different in terms of behavior, usage, needs, and attitudes and their treatment should be tailored to their differentiating characteristics. Clustering techniques attempt to do exactly that: identify distinct customer typologies and segment the customer base into groups of similar profiles so that they can be marketed more effectively.

These techniques automatically detect the underlying customer groups based on an input set of fields/attributes. Clusters are not known in advance. They are revealed by analyzing the observed input data patterns. Clustering techniques assess the similarity of the records or customers with respect to the clustering fields and assign them to the revealed clusters accordingly. The goal is to detect groups with internal homogeneity and interclass heterogeneity.

Clustering techniques are quite popular and their use is widespread in data mining and market research. They can support the development of different segmentation schemes according to the clustering attributes used: namely, behavioral, attitudinal, or demographic segmentation.

The major advantage of the clustering techniques is that they can efficiently manage a large number of attributes and create data-driven segments. The created segments are not based on a priori personal concepts, intuitions, and perceptions of the business people. They are induced by the observed data patterns and, provided they are built properly, they can lead to results with real business meaning and value. Clustering models can analyze complex input data patterns and suggest solutions that would not otherwise be apparent. They reveal customer typologies, enabling tailored marketing strategies. In later chapters we will have the chance to present real-world applications from major industries such as telecommunications and banking, which will highlight the true benefits of data mining-derived clustering solutions.

Unlike classification modeling, in clustering there is no predefined set of classes. There are no predefined categories such as churners/non-churners or buyers/non-buyers and there is also no historical dataset with pre-classified records. It is up to the algorithm to uncover and define the classes and assign each record to its "nearest" or, in other words, its most similar cluster. To present the basic concepts of clustering, let us consider the hypothetical case of a mobile telephony network operator that wants to segment its customers according to their voice and SMS usage. The available demographic data are not used as clustering inputs in this case since the objective concerns the grouping of customers according only to behavioral criteria.

The input dataset, for a few imaginary customers, is presented in Table 2.6.

In the scatterplot in Figure 2.10, these customers are positioned in a two-dimensional space according to their voice usage, along the X-axis, and their SMS usage, along the Y-axis.

The clustering procedure is depicted in Figure 2.11, where voice and SMS usage intensity are represented by the corresponding symbols.

Examination of the scatterplot reveals specific similarities among the customers. Customers 1 and 6 appear close together and present heavy voice usage and low SMS usage. They can be placed in a single group which we label as "Heavy voice users." Similarly, customers 2 and 3 also appear close together but far apart from the rest. They form a group of their own, characterized by average voice and SMS usage. Therefore one more cluster has been disclosed, which can be labeled as "Typical users." Finally, customers 4 and 5 also seem to be different from the

Table 2.6 The modeling dataset for a clustering model.

	Input fields	
Customer ID	**Monthly average number of SMS calls**	**Monthly average number of voice calls**
1	27	144
2	32	44
3	41	30
4	125	21
5	105	23
6	20	121

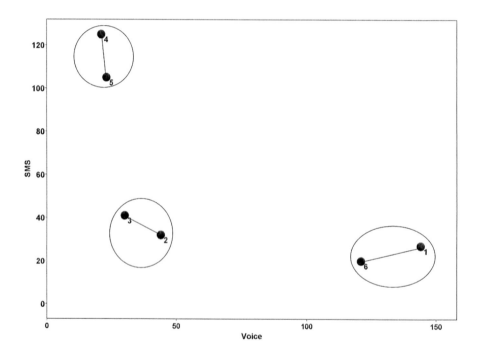

Figure 2.10 Scatterplot of voice and SMS usage.

rest by having increased SMS usage and low voice usage. They can be grouped together to form a cluster of "SMS users."

Although quite naive, the above example outlines the basic concepts of clustering. Clustering solutions are based on analyzing similarities among records. They typically use distance measures that assess the records' similarities and assign

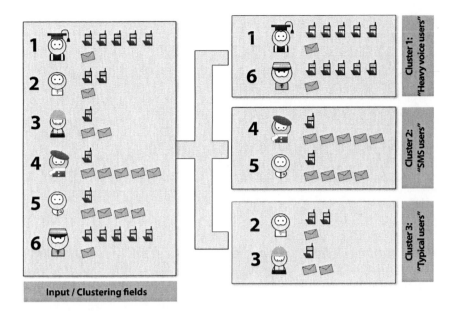

Figure 2.11   Graphical representation of clustering.

records with similar input data patterns, hence similar behavioral profiles, to the same cluster.

Nowadays, various clustering algorithms are available, which differ in their approach for assessing the similarity of records and in the criteria they use to determine the final number of clusters. The whole clustering "revolution" started with a simple and intuitive distance measure, still used by some clustering algorithms today, called the Euclidean distance. The Euclidean distance of two records or objects is a dissimilarity measure calculated as the square root of the sum of the squared differences between the values of the examined attributes/fields. In our example the Euclidean distance between customers 1 and 6 would be:

$$\sqrt{[(\text{Customer 1 voice usage} - \text{Customer 6 voice usage})^2}$$
$$+ (\text{Customer 1 SMS usage} - \text{Customer 6 SMS usage})^2] = 24$$

This value denotes the disparity of customers 1 and 6 and is represented in the respective scatterplot by the length of the straight line that connects points 1 and 6. The Euclidean distances for all pairs of customers are summarized in Table 2.7.

A traditional clustering algorithm, named agglomerative or hierarchical clustering, works by evaluating the Euclidean distances between all pairs of records, literally the length of their connecting lines, and begins to group them accordingly

Table 2.7   The proximity matrix of Euclidean distances between all pairs of customers.

	Euclidean distance					
	1	**2**	**3**	**4**	**5**	**6**
1	0.0	100.1	114.9	157.3	144.0	**24.0**
2	100.1	0.0	**16.6**	95.8	76.0	77.9
3	114.9	16.6	0.0	84.5	64.4	93.4
4	157.3	95.8	84.5	0.0	**20.1**	145.0
5	144.0	76.0	64.4	20.1	0.0	129.7
6	24.0	77.9	93.4	145.0	129.7	0.0

in successive steps. Although many things have changed in clustering algorithms since the inception of this algorithm, it is nice to have a graphical representation of what clustering is all about. Nowadays, in an effort to handle large volumes of data, algorithms use more efficient distance measures and approaches which do not require the calculation of the distances between all pairs of records. Even a specific type of neural network is applied for clustering; however, the main concept is always the same – the grouping of homogeneous records. Typical clustering tasks involve the mining of thousands of records and tens or hundreds of attributes. Things are much more complicated than in our simplified exercise. Tasks like this are impossible to handle without the help of specialized algorithms that aim to automatically uncover the underlying groups.

One thing that should be made crystal clear about clustering is that it *groups records according to the observed input data patterns*. Thus, the data miners and marketers involved should decide in advance, according to the specific business objective, the segmentation level and the segmentation criteria – in other words, the clustering fields. For example, if we want to segment bank customers according to their product balances, we must prepare a modeling dataset with balance information at a customer level. Even if our original input data are in a transactional format or stored at a product account level, the selected segmentation level requires a modeling dataset with a unique record per customer and with fields that would summarize their product balances.

In general, clustering algorithms provide an exhaustive and mutual exclusive solution. They automatically assign each record to one of the uncovered groups. They produce disjoint clusters and generate a cluster membership field that denotes the group of each record, as shown in Table 2.8.

In our illustrative exercise we have discovered the differentiating characteristics of each cluster and labeled them accordingly. In practice, this process is not so easy and may involve many different attributes, even those not directly participating

Table 2.8   The cluster membership field.

	Input fields		Model-generated field
Customer ID	Average monthly number of SMS calls	Average monthly number of voice calls	Cluster membership field
1	27	144	**Cluster 1 – heavy voice users**
2	32	44	**Cluster 2 – typical users**
3	41	30	**Cluster 2 – typical users**
4	125	21	**Cluster 2 – SMS users**
5	105	23	**Cluster 2 – SMS users**
6	20	121	**Cluster 1 – heavy voice users**

in the cluster formation. Each clustering solution should be thoroughly examined and the profiles of the clusters outlined. This is usually accomplished by simple reporting techniques, but it can also include the application of supervised modeling techniques such as classification techniques, aiming to reveal the distinct characteristics associated with each cluster.

This profiling phase is an essential step in the clustering procedure. It can provide insight on the derived segmentation scheme and it can also help in the evaluation of the scheme's usefulness. The derived clusters should be evaluated with respect to the business objective they were built to serve. The results should make sense from a business point of view and should generate business opportunities. The marketers and data miners involved should try to evaluate different solutions before selecting the one that best addresses the original business goal.

Available clustering models include the following:

• **Agglomerative or hierarchical:** Although quite outdated nowadays, we present this algorithm since in a way it is the "mother" of all clustering models. It is called hierarchical or agglomerative because it starts with a solution where each record comprises a cluster and gradually groups records up to the point where all of them fall into one supercluster. In each step it calculates the distances between all pairs of records and groups the most similar ones. A table (agglomeration schedule) or a graph (dendrogram) summarizes the grouping steps and the respective distances. The analyst should consult this information, identify the point where the algorithm starts to group disjoint cases, and then decide on the number of clusters to retain. This algorithm cannot effectively handle more than a few thousand cases. Thus it cannot be directly applied in most business clustering tasks. A usual workaround is to a use it on a sample of the clustering population. However, with numerous other efficient algorithms that can easily handle millions of records, clustering through sampling is not considered an ideal approach.

- **K-means:** This is an efficient and perhaps the fastest clustering algorithm that can handle both long (many records) and wide datasets (many data dimensions and input fields). It is a distance-based clustering technique and, unlike the hierarchical algorithm, it does not need to calculate the distances between all pairs of records. The number of clusters to be formed is predetermined and specified by the user in advance. Usually a number of different solutions should be tried and evaluated before approving the most appropriate. It is best for handling continuous clustering fields.
- **TwoStep cluster:** As its name implies, this scalable and efficient clustering model, included in IBM™ SPSS™ Modeler (formerly Clementine), processes records in two steps. The first step of pre-clustering makes a single pass through the data and assigns records to a limited set of initial subclusters. In the second step, initial subclusters are further grouped, through hierarchical clustering, into the final segments. It suggests a clustering solution by automatic clustering: the optimal number of clusters can be automatically determined by the algorithm according to specific criteria.
- **Kohonen network/Self-Organizing Map (SOM):** Kohonen networks are based on neural networks and typically produce a two-dimensional grid or map of the clusters, hence the name self-organizing maps. Kohonen networks usually take a longer time to train than the K-means and TwoStep algorithms, but they provide a different view on clustering that is worth trying.

Apart from segmentation, clustering techniques can also be used for other purposes, for example, as a preparatory step for optimizing the results of predictive models. Homogeneous customer groups can be revealed by clustering and then separate, more targeted predictive models can be built within each cluster. Alternatively, the derived cluster membership field can also be included in the list of predictors in a supervised model. Since the cluster field combines information from many other fields, it often has significant predictive power. Another application of clustering is in the identification of unusual records. Small or outlier clusters could contain records with increased significance that are worth closer inspection. Similarly, records far apart from the majority of the cluster members might also indicate anomalous cases that require special attention.

The clustering techniques are further explained and presented in detail in the next chapter.

REDUCING THE DIMENSIONALITY OF DATA WITH DATA REDUCTION TECHNIQUES

As their name implies, data reduction techniques aim at effectively reducing the data's dimensions and removing redundant information. They do so by replacing the initial set of fields with a core set of compound measures which simplify subsequent modeling while retaining most of the information of the original attributes.

Factor analysis and PCA are among the most popular data reduction techniques. They are unsupervised, statistical techniques which deal with continuous input attributes. These attributes are analyzed and mapped to representative fields, named factors or components. The procedure is illustrated in Figure 2.12.

Factor analysis and PCA are based on the concept of linear correlation. If certain continuous fields/attributes tend to covary then they are correlated. If their relationship is expressed adequately by a straight line then they have a strong linear correlation. The scatterplot in Figure 2.13 depicts the monthly average SMS and MMS (Multimedia Messaging Service) usage for a group of mobile telephony customers.

As seen in the scatterplot, most customer points cluster around a straight line with a positive slope that slants upward to the right. Customers with increased SMS

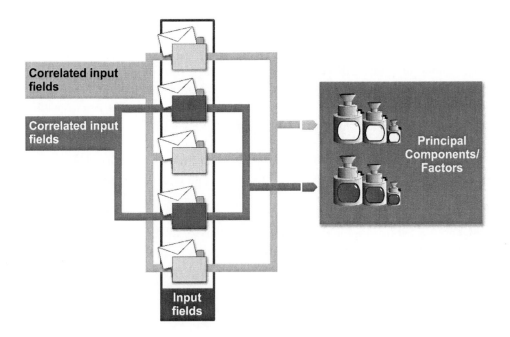

Figure 2.12   Data reduction techniques.

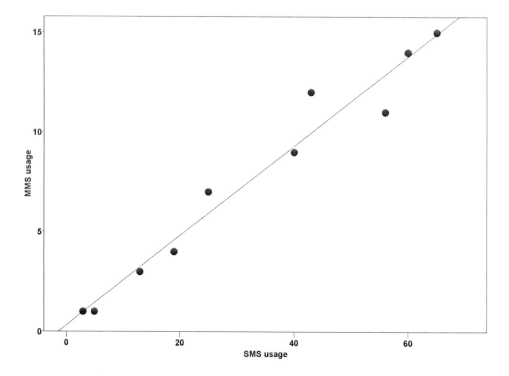

Figure 2.13   Linear correlation between two continuous measures.

usage also tend to be MMS users as well. These two services are related in a linear manner and present a strong, positive linear correlation, since high values of one field tend to correspond to high values of the other. However, in negative linear correlations, the direction of the relationship is reversed. These relationships are described by straight lines with a negative slope that slant downward. In such cases high values of one field tend to correspond to low values of the other. The strength of linear correlation is quantified by a measure named the Pearson correlation coefficient. It ranges from −1 to +1. The sign of the coefficient reveals the direction of the relationship. Values close to +1 denote strong positive correlation and values close to −1 negative correlation. Values around 0 denote no discernible linear correlation, yet this does not exclude the possibility of nonlinear correlation.

Factor analysis and PCA examine the correlations between the original input fields and identify latent data dimensions. In a way they "group" the inputs into composite measures, named factors or components, that can effectively represent the original attributes, without sacrificing much of their information. The derived components and factors have the form of continuous numeric scores and can be subsequently used as any other fields for reporting or modeling purposes.

Data reduction is also widely used in marketing research. The views, perceptions, and preferences of the respondents are often recorded through a large number of questions that investigate all the topics of interest in detail. These questions often have the form of a Likert scale, where respondents are asked to state, on a scale of 1–5, the degree of importance, preference, or agreement on specific issues. The answers can be used to identify the latent concepts that underlie the respondents' views.

To further explain the basic concepts behind data reduction techniques, let us consider the simple case of a few customers of a mobile telephony operator. SMS, MMS, and voice call traffic, specifically the number of calls by service type and the minutes of voice calls, were analyzed by principal components. The modeling dataset and the respective results are given in Table 2.9.

The PCA model analyzed the associations among the original fields and identified two components. More specifically, the SMS and MMS usage appear to be correlated and a new component was extracted to represent the usage of those services. Similarly, the number and minutes of voice calls were also correlated. The second component represents these two fields and measures the voice usage intensity. Each derived component is standardized, with an overall population mean of 0 and a standard deviation of 1. The component scores denote how many standard deviations above or below the overall mean each record stands. In simple terms, a positive score in component 1 indicates high SMS and MMS usage while a negative score indicates below-average usage. Similarly, high scores on component

Table 2.9 The modeling dataset for principal components analysis and the derived component scores.

	Input fields				Model-generated fields	
Customer ID	Monthly average number of SMS calls	Monthly average number of MMS calls	Monthly average number of voice calls	Monthly average number of voice call minutes	Component 1 score – "SMS/MMS usage"	Component 2 score – "voice usage"
1	19	4	90	150	−0.57	1.99
2	43	12	30	35	0.61	−0.42
3	13	3	10	20	−0.94	−1.05
4	60	14	100	80	1.34	1.38
5	5	1	30	55	−1.27	−0.29
6	56	11	25	35	0.78	−0.48
7	25	7	30	28	−0.25	−0.57
8	3	1	65	82	−1.23	0.65
9	40	9	15	30	0.22	−0.76
10	65	15	20	40	1.33	−0.46

2 denote high voice usage, in terms of both frequency and duration of calls. The generated scores can then be used in subsequent modeling tasks.

The interpretation of the derived components is an essential part of the data reduction procedure. Since the derived components will be used in subsequent tasks, it is important to fully understand the information they convey. Although there are many formal criteria for selecting the number of factors to be retained, analysts should also examine their business meaning and only keep those that comprise interpretable and meaningful measures.

Simplicity is the key benefit of data reduction techniques, since they drastically reduce the number of fields under study to a core set of composite measures. Some data mining techniques may run too slow or not at all if they have to handle a large number of inputs. Situations like these can be avoided by using the derived component scores instead of the original fields. An additional advantage of data reduction techniques is that they can produce uncorrelated components. This is one of the main reasons for applying a data reduction technique as a preparatory step before other models. Many predictive modeling techniques can suffer from the inclusion of correlated predictors, a problem referred to as multicollinearity. By substituting the correlated predictors with the extracted components we can eliminate collinearity and substantially improve the stability of the predictive model. Additionally, clustering solutions can also be biased if the inputs are dominated by correlated "variants" of the same attribute. By using a data reduction technique we can unveil the true data dimensions and ensure that they are of equal weight in the formation of the final clusters.

In the next chapter, we will revisit data reduction techniques and present PCA in detail.

FINDING "WHAT GOES WITH WHAT" WITH ASSOCIATION OR AFFINITY MODELING TECHNIQUES

When browsing a bookstore on the Internet you may have noticed recommendations that pop up and suggest additional, related products for you to consider: "Customers who have bought this book have also bought the following books." Most of the time these recommendations are quite helpful, since they take into account the recorded preferences of past customers. Usually they are based on association or affinity data mining models.

These models analyze past co-occurrences of events, purchases, or attributes and detect associations. They associate a particular outcome category, for instance a product, with a set of conditions, for instance a set of other products. They are typically used to identify purchase patterns and groups of products purchased together.

In the e-bookstore example, by browsing through past purchases, association models can discover other popular books among the buyers of the particular book viewed. They can then generate individualized recommendations that match the indicated preference.

Association modeling techniques generate rules of the following general format:

```
IF (ANTECEDENTS) THEN CONSEQUENT
```

For example:

```
IF (product A and product C and product E and ...) → product B
```

More specifically, a rule referring to supermarket purchases might be:

```
IF EGGS & MILK & FRESH FRUIT → VEGETABLES
```

This simple rule, derived by analyzing past shopping carts, identifies associated products that tend to be purchased together: when eggs, milk, and fresh fruit are bought, then there is an increased probability of also buying vegetables. This probability, referred to as the rule's confidence, denotes the rule's strength and will be further explained in what follows.

The left or the IF part of the rule consists of the **antecedent** or condition: a situation where, when true, the rule applies and the consequent shows increased occurrence rates. In other words, the antecedent part contains the product combinations that usually lead to some other product. The right part of the rule is the **consequent** or conclusion: what tends to be true when the antecedents hold true. The rule's complexity depends on the number of antecedents linked to the consequent.

These models aim at:

- **Providing insight on product affinities:** Understand which products are commonly purchased together. This, for instance, can provide valuable information for advertising, for effectively reorganizing shelves or catalogues, and for developing special offers for bundles of products or services.
- **Providing product suggestions:** Association rules can act as a recommendation engine. They can analyze shopping carts and help in direct marketing activities by producing personalized product suggestions, according to the customer's recorded behavior.

This type of analysis is also referred to as **market basket analysis** since it originated from point-of-sale data and the need to understand consumer shopping patterns. Its application was extended to also cover any other "basket-like" problem from various other industries. For example:

- In banking, it can be used for finding common product combinations owned by customers.
- In telecommunications, for revealing services that usually go together.
- In web analysis, for finding web pages accessed in single visits.

Association models are considered as unsupervised techniques since they do not involve a single output field to be predicted. They analyze product affinity tables: that is, multiple fields that denote product/service possession. These fields are at the same time considered as inputs and outputs. Thus, all products are predicted and act as predictors for the rest of the products.

According to the business scope and the selected level of analysis, association models can be applied to:

- Transaction or order data – data summarizing purchases at a transaction level, for instance what is bought in a single store visit.
- Aggregated information at a customer level – what is bought during a specific time period by each customer or what is the current product mix of each (bank) customer.

Product Groupings

In general, these techniques are rarely applied directly to product codes. They are usually applied to product groups. A taxonomy level, also referred to as a hierarchy or grouping level, is selected according to the defined business objective and the data are grouped accordingly. The selected product group- ing will also determine the type of generated rules and recommendations.

A typical modeling dataset for an association model has the tabular format shown in Table 2.10. These tables, also known as basket or truth tables, contain categorical, flag (binary) fields which denote the presence or absence of specific items or events of interest, for instance purchased products. The fields denoting product purchases, or in general event occurrences, are the content fields. The analysis ID field, here the transaction ID, is used to define the unit or level of the analysis. In other words, whether the revealed purchase patterns refer to transactions or customers. In tabular data format, the dataset should contain aggregated content/purchase information at the selected analysis level.

Table 2.10 The modeling dataset for association modeling – a basket table.

| Analysis ID field | Input–output fields | | | |
| | Content fields | | | |
Transaction ID	Product 1	Product 2	Product 3	Product 4
101	True	False	True	False
102	True	False	False	False
103	True	False	True	True
104	True	False	False	True
105	True	False	False	True
106	False	True	True	False
107	True	False	True	True
108	False	False	True	True
109	True	False	True	True

In the above example, the goal is to analyze purchase transactions and identify rules which describe the shopping cart patterns. We also assume that products are grouped into four supercategories.

Analyzing Raw Transactional Data with Association Models

Besides basket tables, specific algorithms, like the a priori association model, can also directly analyze transactional input data. This format requires the presence of two fields: a content field denoting the associated items and an analysis ID field that defines the level of analysis. Multiple records are linked by having the same ID value. The transactional modeling dataset for the simple example presented above is listed in Table 2.11.

Table 2.11 A transactional modeling dataset for association modeling.

| Analysis ID field | Input–output field |
| | Content field |
Transaction ID	Products
101	Product 1
101	Product 3

Table 2.11   *(continued)*

Input–output field	
Analysis ID field	**Content field**
Transaction ID	**Products**
102	Product 2
103	Product 1
103	Product 3
103	Product 4
104	Product 1
104	Product 4
105	Product 1
105	Product 4
106	Product 2
106	Product 3
107	Product 1
107	Product 3
107	Product 4
108	Product 3
108	Product 4
109	Product 1
109	Product 3
109	Product 4

By setting the transaction ID field as the analysis ID we require the algorithm to analyze the purchase patterns at a transaction level. If the customer ID had been selected as the analysis ID, the purchase transactions would have been internally aggregated and analyzed at a customer level.

Two of the derived rules are listed in Table 2.12.

Usually all the extracted rules are described and evaluated with respect to three main measures:

Table 2.12   Rules of an association model.

Rule ID	Consequent	Antecedent	Support %	Confidence %	Lift
Rule 1	Product 4	Products 3 and 1	44.4	75.0	1.13
Rule 2	Product 4	Product 1	77.8	71.4	1.07

- **The support:** This assesses the rule's coverage or "how many records the rule constitutes." It denotes the percentage of records that match the antecedents.
- **The confidence:** This assesses the strength and the predictive ability of the rule. It indicates "how likely the consequent is, given the antecedents." It denotes the consequent percentage or probability, within the records that match the antecedents.
- **The lift:** This assesses the improvement in the predictive ability when using the derived rule compared to randomness. It is defined as the ratio of the rule confidence to the prior confidence of the consequent. The prior confidence is the overall percentage of the consequent within all the analyzed records.

In the presented example, Rule 2 associates product 1 to product 4 with a confidence of 71.4%. In plain English, it states that 71.4% of the baskets containing product 1, which is the antecedent, also contain product 4, the consequent. Additionally, the baskets containing product 1 comprise 77.8% of all the baskets analyzed. This measure is the support of the rule. Since six out of the nine total baskets contain product 4, the prior confidence of a basket containing product 4 is 6/9 or 67%, slightly lower than the rule confidence. Specifically, Rule 2 outperforms randomness and achieves a confidence about 7% higher with a lift of 1.07. Thus by using the rule, the chances of correctly identifying a product 1 purchase are improved by 7%.

Rule 4 is more complicated since it contains two antecedents. It has a lower coverage (44.4%) but yields a higher confidence (75%) and lift (1.13). In plain English this rule states that baskets with products 1 and 3 present a strong chance (75%) of also containing product 4. Thus, there is a business opportunity to promote product 4 to all customers who check out with products 1 and 3 and have not bought product 4.

The rule development procedure can be controlled according to model parameters that analysts can specify. Specifically, analysts can define in advance the required threshold values for rule complexity, support, confidence, and lift in order to guide the rule growth process according to their specific requirements.

Unlike decision trees, association models generate rules that overlap. Therefore, multiple rules may apply for each customer. Rules applicable to each customer are then sorted according to a selected performance measure, for instance lift or confidence, and a specified number of *n* rules, for instance the top three rules, are retained. The retained rules indicate the top *n* product suggestions, currently not in the basket, that best match each customer's profile. In this way, association models can help in cross-selling activities as they can provide specialized product recommendations for each customer. As in every data mining task, derived rules should also be evaluated with respect to their business meaning and "actionability" before deployment.

Association vs. Classification Models for Product Suggestions

As described above, the association models can be used for cross selling and for identification of the next best offers for each customer. Although useful, association modeling is not the ideal approach for next best product campaigns, mainly because they do not take into account customer evolvement and possible changes in the product mix over time.

A recommended approach would be to analyze the profile of customers before the uptake of a product to identify the characteristics that have caused the event and are not the result of the event. This approach is feasible by using either test campaign data or historical data. For instance, an organization might conduct a pilot campaign among a sample of customers not owning a specific product that it wants to promote, and mine the collected results to identify the profile of customers most likely to respond to the product offer. Alternatively, it can use historical data, and analyze the profile of those customers who recently acquired the specific product. Both these approaches require the application of a classification model to effectively estimate acquisition propensities.

Therefore, a set of separate classification models for each product and a procedure that would combine the estimated propensities into a next best offer strategy are a more efficient approach than a set of association rules.

Most association models include categorical and specifically binary (flag or dichotomous) fields, which typically denote product possession or purchase. We can also include supplementary fields, like demographics, in order to enhance the antecedent part of the rules and enrich the results. These fields must also be categorical, although specific algorithms, like GRI (Generalized Rule Induction), can also handle continuous supplementary fields. The a priori algorithm is perhaps the most widely used association modeling technique.

DISCOVERING EVENT SEQUENCES WITH SEQUENCE MODELING TECHNIQUES

Sequence modeling techniques are used to identify associations of events/purchases/attributes over time. They take into account the order of events and detect sequential associations that lead to specific outcomes. They generate rules analogous to association models but with one difference: a sequence of antecedent events is strongly associated with the occurrence of a consequent. In other words, when certain things happen in a specific order, a specific event has an increased probability of occurring next.

Sequence modeling techniques analyze paths of events in order to detect common sequences. Their origin lies in web mining and click stream analysis of web pages. They began as a way to analyze weblog data in order to understand the navigation patterns in web sites and identify the browsing trails that end up in specific pages, for instance purchase checkout pages. The use of these techniques has been extended and nowadays can be applied to all "sequence" business problems.

The techniques can also be used as a means for predicting the next expected "move" of the customers or the next phase in a customer's lifecycle. In banking, they can be applied to identify a series of events or customer interactions that may be associated with discontinuing the use of a credit card; in telecommunications, for identifying typical purchase paths that are highly associated with the purchase of a particular add-on service; and in manufacturing and quality control, for uncovering signs in the production process that lead to defective products.

The rules generated by association models include antecedents or conditions and a consequent or conclusion. When antecedents occur in a specific order, it is likely that they will be followed by the occurrence of the consequent. Their general format is:

```
IF (ANTECEDENTS with a specific order) THEN CONSEQUENT
```

or:

```
IF (product A and THEN product F and THEN product C and THEN ...) THEN
product D
```

For example, a rule referring to bank products might be:

```
IF SAVINGS & THEN CREDIT CARD & THEN SHORT-TERM DEPOSIT → STOCKS
```

This rule states that bank customers who start their relationship with the bank as savings account customers, and subsequently acquire a credit card and a short-term deposit product, present an increased likelihood to invest in stocks. The likelihood of the consequent, given the antecedents, is expressed by the confidence value. The confidence value assesses the rule's strength. Support and confidence measures, which were presented in detail for association models, are also applicable in sequence models.

The generated sequence models, when used for scoring, provide a set of predictions denoting the n, for instance the three, most likely next steps, given the observed antecedents. <u>Predictions are sorted in terms of their confidence and may indicate for example the top three next product suggestions for each customer according to his or her recorded path of product purchasing to date.</u>

Sequence models require the presence of an ID field to monitor the events of the same individual over time. The sequence data could be tabular or transactional, in a format similar to the one presented for association modeling. Fields required for the analysis involve: content(s) field(s), an analysis ID field, and a time field. Content(s) fields denote the occurrence of events of interest, for instance purchased products or web pages viewed during a visit to a site. <u>The analysis ID field determines the level of analysis, for instance whether the revealed sequence patterns would refer to customers, transactions, or web visits, based on appropriately prepared weblog files.</u> The time field records the time of the events and is required so that the algorithm can track the occurrence order. A typical transactional modeling dataset, recording customer purchases over time, is given in Table 2.13.

A derived association rule is displayed in Table 2.14.

Table 2.13   A transactional modeling data set for association modeling.

		Input–output field
Analysis ID field	**Time field**	**Content field**
Customer ID	**Acquisition time**	**Products**
101	30 June 2007	Product 1
101	12 August 2007	Product 3
101	20 December 2008	Product 4
102	10 September 2008	Product 3
102	12 September 2008	Product 5
102	20 January 2009	Product 5
103	30 January 2009	Product 1
104	10 January 2009	Product 1
104	10 January 2009	Product 3
104	10 January 2009	Product 4
105	10 January 2009	Product 1
105	10 February 2009	Product 5
106	30 June 2007	Product 1
106	12 August 2007	Product 3
106	20 December 2008	Product 4
107	30 June 2007	Product 2
107	12 August 2007	Product 1
107	20 December 2008	Product 3

Table 2.14   Rule of an association detection model.

Rule ID	Consequent	Antecedents	Support %	Confidence %
Rule 1	Product 4	Product 1 then Product 3 (4/7).	57.1	75.0

The support value represents the percentage of units of the analysis, here unique customers, that had a sequence of the antecedents. In the above example the support rises to 57.1%, since four out of seven customers purchased product 3 after buying product 1. Three of these customers purchased product 4 afterward. Thus, the respective rule confidence figure is 75%. The rule simply states that after acquiring product 1 and then product 3, customers have an increased likelihood (75%) of purchasing product 4 next.

DETECTING UNUSUAL RECORDS WITH RECORD SCREENING MODELING TECHNIQUES

Record screening modeling techniques are applied to detect anomalies or outliers. The techniques try to identify records with odd data patterns that do not "conform" to the typical patterns of "normal" cases.

Unsupervised record screening modeling techniques can be used for:

• Data auditing, as a preparatory step before applying subsequent data mining models.
• Discovering fraud.

Valuable information is not just hidden in general data patterns. Sometimes rare or unexpected data patterns can reveal situations that merit special attention or require immediate action. For instance, in the insurance industry, unusual claim profiles may indicate fraudulent cases. Similarly, odd money transfer transactions may suggest money laundering. Credit card transactions that do no fit the general usage profile of the owner may also indicate signs of suspicious activity.

Record screening modeling techniques can provide valuable help in revealing fraud by identifying "unexpected" data patterns and "odd" cases. The unexpected cases are not always suspicious. They may just indicate an unusual, yet acceptable, behavior. For sure, though, they require further investigation before being classified as suspicious or not.

Record screening models can also play another important role. They can be used as a data exploration tool before the development of another data mining model. Some models, especially those with a statistical origin, can be affected by the presence of abnormal cases which may lead to poor or biased solutions. It is always a good idea to identify these cases in advance and thoroughly examine them before deciding on their inclusion in subsequent analysis.

Modified standard data mining techniques, like clustering models, can be used for the unsupervised detection of anomalies. Outliers can often be found among cases that do not fit well in any of the emerged clusters or in sparsely populated clusters. Thus, the usual tactic for uncovering anomalous records is to develop an explorative clustering solution and then further investigate the results. A specialized technique in the field of unsupervised record screening is IBM SPSS Modeler's Anomaly Detection. It is an exploratory technique based on clustering. It provides a quick, preliminary data investigation and suggests a list of records with odd data patterns for further investigation. It evaluates each record's "normality" in a multivariate context and not on a per-field base by assessing all the inputs together. More specifically, it identifies peculiar cases by deriving a cluster solution and then measuring the distance of each record from its cluster central point, the centroid. An anomaly index is then calculated that represents the proximity of each record to the other records in its cluster. Records can be sorted according to this measure and then flagged as anomalous according to a user-specified threshold value. What is interesting about this algorithm is that it provides the reasoning behind its results. For each anomalous case it displays the fields with the unexpected values that do not conform to the general profile of the record.

Supervised and Unsupervised Models for Detecting Fraud

Unsupervised record screening techniques can be applied for fraud detection even in the absence of recorded fraudulent events. If past fraudulent cases are available, analysts can try a supervised classification model to identify the input data patterns associated with the target suspicious activities. The supervised approach has strengths since it works in a more targeted way than unsupervised record screening. However, it also has specific disadvantages. Since fraudsters' behaviors may change and evolve over time, supervised models trained on past cases may soon become outdated and fail to capture new tricks and new types of suspicious patterns. Additionally, the list of past fraudulent cases, which is necessary for the training of the classification model, is often biased and partial. It depends on the specific rules and criteria in use. The existing list may not cover all types of potential fraud and may need to be appended to the results of random audits. In conclusion, both the supervised and unsupervised approaches for detecting fraud have pros and cons. A combined approach is the one that usually yields the best results.

MACHINE LEARNING/ARTIFICIAL INTELLIGENCE VS. STATISTICAL TECHNIQUES

According to their origin and the way they analyze data patterns, the data mining models can be grouped into two classes:

- Machine learning/artificial intelligence models
- Statistical models.

Statistical models include algorithms like OLSR, logistic regression, factor analysis/PCA, among others. Techniques like decision trees, neural networks, association rules, self-organizing maps are machine learning models.

With the rapid developments in IT in recent years, there has been a rapid growth in machine leaning algorithms, expanding analytical capabilities in terms of both efficiency and scalability. Nevertheless, one should never underestimate the predictive power of "traditional" statistical techniques whose robustness and reliability have been established and proven over the years.

Faced with the growing volume of stored data, analysts started to look for faster algorithms that could overcome potential time and size limitations. Machine learning models were developed as an answer to the need to analyze large amounts of data in a reasonable time. New algorithms were also developed to overcome certain assumptions and restrictions of statistical models and to provide solutions to newly arisen business problems like the need to analyze affinities through association modeling and sequences through sequence models.

Trying many different modeling solutions is the essence of data mining. There is no particular technique or class of techniques which yields superior results in all situations and for all types of problems. However, in general, machine learning algorithms perform better than traditional statistical techniques in regard to speed and capacity of analyzing large volumes of data. Some traditional statistical techniques may fail to efficiently handle wide (high-dimensional) or long datasets (many records). For instance, in the case of a classification project, a logistic regression model would demand more resources and processing time than a decision tree model. Similarly, a hierarchical or agglomerative cluster algorithm will fail to analyze more than a few thousand records when some of the most recently developed clustering algorithms, like IBM SPSS Modeler TwoStep Model, can handle millions without sampling. Within the machine learning algorithms we can also note substantial differences in terms of speed and required resources, with neural networks, including SOMs for clustering, among the most demanding techniques.

Another advantage of machine learning algorithms is that they have less stringent data assumptions. Thus they are more friendly and simple to use for those with little experience in the technical aspects of model building. Usually, statistical algorithms require considerable effort in building. Analysts should spend time taking into account the data considerations. Merely feeding raw data into these algorithms will probably yield poor results. Their building may require special data processing and transformations before they produce results comparable or even superior to those of machine learning algorithms.

Another aspect that data miners should take into account when choosing a model technique is the insight provided by each algorithm. In general, statistical models yield transparent solutions. On the contrary, some machine learning models, like neural networks, are opaque, conveying little information and knowledge about the underlying data patterns and customer behaviors. They may provide reliable customer scores and achieve satisfactory predictive performance, but they provide little or no reasoning for their predictions. However, among machine learning algorithms there are models that provide an explanation of the derived results, like decision trees. Their results are presented in an intuitive and self-explanatory format, allowing an understanding of the findings. Since most data mining software packages allow for fast and easy model development, the case of developing one model for insight and a different model for scoring and deployment is not unusual.

SUMMARY

In the previous sections we presented a brief introduction to the main concepts of data mining modeling techniques. Models can be grouped into two main classes: supervised and unsupervised.

Supervised modeling techniques are also referred to as directed or predictive because their goal is prediction. Models automatically detect or "learn" the input data patterns associated with specific output values. Supervised models are further grouped into classification and estimation models, according to the measurement level of the target field. Classification models deal with the prediction of categorical outcomes. Their goal is to classify new cases into predefined classes. Classification models can support many important marketing applications that are related to the prediction of events, such as customer acquisition, cross/up/deep selling, and churn prevention. These models estimate event scores or propensities for all the examined customers, which enable marketers to efficiently target their subsequent campaigns and prioritize their actions. Estimation models, on the other hand, aim at estimating the values of continuous target fields. Supervised models require a thorough evaluation of their performance before deployment. There are many evaluation tools and methods which mainly include the cross-examination of the model's predicted results with the observed actual values of the target field.

In Table 2.15 we present a list summarizing the supervised modeling techniques, in the fields of classification and estimation. The table is not meant to be exhaustive but rather an indicative listing which presents some of the most well-known and established algorithms.

While supervised models aim at prediction, unsupervised models are mainly used for grouping records or fields and for the detection of events or attributes that occur together. Data reduction techniques are used to narrow the data's dimensions, especially in the case of wide datasets with correlated inputs. They identify related sets of original fields and derive compound measures that can effectively replace them in subsequent tasks. They simplify subsequent modeling or reporting jobs without sacrificing much of the information contained in the initial list of fields.

Table 2.15   Supervised modeling techniques.

Classification techniques	Estimation/regression techniques
• Logistic regression • Decision trees: • C5.0 • CHAID • Classification and Regression Trees • QUEST • Decision rules: • C5.0 • Decision list • Discriminant analysis • Neural networks • Support vector machine • Bayesian networks	• Ordinary least squares regression • Neural networks • Decision trees: • CHAID • Classification and Regression Trees • Support vector machine • Generalized linear models

Table 2.16 Unsupervised modeling techniques.

Clustering techniques	Data reduction techniques
• K-means • TwoStep cluster • Kohonen network/self-organizing map	• Principal components analysis • Factor analysis

Clustering models automatically detect natural groupings of records. They can be used to segment customers. All customers are assigned to one of the derived clusters according to their input data patterns and their profiles. Although an explorative technique, clustering also requires the evaluation of the derived clusters before selecting a final solution. The revealed clusters should be understandable, meaningful, and actionable in order to support the development of an effective segmentation scheme.

Association models identify events/products/attributes that tend to co-occur. They can be used for market basket analysis and in all other "affinity" business problems related to questions such as "what goes with what?" They generate IF . . . THEN rules which associate antecedents to a specific consequent. Sequence models are an extension of association models that also take into account the order of events. They detect sequences of events and can be used in web path analysis and in any other "sequence" type of problem.

Table 2.16 lists unsupervised modeling techniques in the fields of clustering and data reduction. Once again the table is not meant to be exhaustive but rather an indicative listing of some of the most popular algorithms.

One last thing to note about data mining models: they should not be viewed as a stand-alone procedure but rather as one of the steps in a well-designed procedure. Model results depend greatly on the preceding steps of the process (business understanding, data understanding, and data preparation) and on decisions and actions that precede the actual model training. Although most data mining models automatically detect patterns, they also depend on the skills of the persons involved. Technical skills are not enough. They should be complemented with business expertise in order to yield meaningful instead of trivial or ambiguous results. Finally, a model can only be considered as effective if its results, after being evaluated as useful, are deployed and integrated into the organization's everyday business operations.

Since the book focuses on customer segmentation, a thorough presentation of supervised algorithms is beyond its scope. In the next chapter we will introduce only the key concepts of decision trees, as this is a technique that is often used in the framework of a segmentation project for scoring and profiling. We will, however, present in detail in that chapter those data reduction and clustering techniques that are widely used in segmentation applications.

CHAPTER THREE

Data Mining Techniques for Segmentation

SEGMENTING CUSTOMERS WITH DATA MINING TECHNIQUES

In this chapter we focus on the data mining modeling techniques used for segmentation. We will present in detail some of the most popular and efficient clustering algorithms, their settings, strengths, and capabilities, and we will see them in action through a simple example that aims at preparing readers for the real-world applications to be presented in subsequent chapters.

Although clustering algorithms can be directly applied to input data, a recommended preprocessing step is the application of a data reduction technique that can simplify and enhance the segmentation process by removing redundant information. This approach, although optional, is highly recommended, as it adjusts for possible input data intercorrelations, ensuring rich and unbiased segmentation solutions that equally account for all the underlying data dimensions. Therefore, this chapter also presents in detail principal components analysis (PCA), an established data reduction technique typically used for grouping the original fields into meaningful components.

PRINCIPAL COMPONENTS ANALYSIS

PCA is a statistical technique used to reduce the data of the original input fields. It derives a limited number of compound measures that can efficiently substitute for the original inputs while retaining most of their information.

Data Mining Techniques in CRM: Inside Customer Segmentation   K. Tsiptsis and A. Chorianopoulos
© 2009 John Wiley & Sons, Ltd

PCA is based on linear correlations. The concept of linear correlation and the measure of the Pearson correlation coefficient were presented in the previous chapter. PCA examines the correlations among the original inputs and uses this information to construct the appropriate composite measures, named principal components.

The goal of PCA is to extract the smallest number of components which account for as much as possible of the information of the original fields. Moreover, a typical PCA derives uncorrelated components, a characteristic that makes them appropriate as input to many other modeling techniques, including clustering. The derived components are typically associated with a specific set of the original fields. They are produced by linear transformations of the inputs, as shown by the following equations, where F_i denotes the input fields (n fields) used for the construction of the components (m components):

$$\text{Component } 1 = a_{11}{}^*F_1 + a_{12}{}^*F_2 + \cdots + a_{1n}{}^*F_n$$
$$\text{Component } 2 = a_{21}{}^*F_1 + a_{22}{}^*F_2 + \cdots + a_{2n}{}^*F_n$$
$$\vdots$$
$$\text{Component } m = a_{m1}{}^*F_1 + a_{m2}{}^*F_2 + \cdots + a_{mn}{}^*F_n$$

The coefficients are automatically calculated by the algorithm so that the loss of information is minimal. Components are extracted in decreasing order of importance, with the first one being the most significant as it accounts for the largest amount of the total original information. Specifically, the first component is the linear combination that carries as much as possible of the total variability of the input fields. Thus, it explains most of their information. The second component accounts for the largest amount of the unexplained variability and is also uncorrelated with the first component. Subsequent components are constructed to account for the remaining information.

Since n components are required to fully account for the original information of n input fields, the question is "where do we stop and how many factors should we extract?" Although there are specific technical criteria that can be applied to guide analysts in the procedure, the final decision should take into account criteria such as the interpretability and the business meaning of the components. The final solution should balance simplicity with effectiveness, consisting of a reduced and interpretable set of components that can adequately represent the original fields.

Apart from PCA, a related statistical technique commonly used for data reduction is factor analysis. It is a quite similar technique that tends to produce results comparable to PCA. Factor analysis is mostly used when the main scope of the analysis is to uncover and interpret latent data dimensions, whereas PCA is typically the preferred option for reducing the dimensionality of the data.

In the following sections we will focus on PCA. We will examine and explain the PCA results and present guidelines for setting up, understanding, and using this modeling technique. Key issues that a data miner has to face in PCA include:

- How many components are to be extracted?
- Is the derived solution efficient and useful?
- Which original fields are mostly related with each component?
- What does each component represent? In other words, what is the meaning of each component?

The next sections will try to clarify these issues.

PCA DATA CONSIDERATIONS

PCA, as an unsupervised technique, expects only inputs. Specifically, it is appropriate for the analysis of numeric continuous fields. Categorical data are not suitable for this type of analysis.

Moreover, it is assumed that there are linear correlations among at least some of the original fields. Obviously data reduction makes sense only in the case of associated inputs, otherwise the respective benefits are trivial.

Unlike clustering techniques, PCA is not affected by potential differences in the measurement scale of the inputs. Consequently there is no need to compensate for fields measured in larger values than others.

PCA scores new records by deriving new fields representing the component scores, but it will not score incomplete records (records with null or missing values in any of the input fields).

HOW MANY COMPONENTS ARE TO BE EXTRACTED?

In the next section we will present PCA by examining the results of a simple example referring to the case of a mobile telephony operator that wants to analyze customer behaviors and reveal the true data dimensions which underlie the usage fields given in Table 3.1. (Hereafter, for readability in all tables and graphs of results, the field names will be presented without underlines.)

Table 3.2 lists the pairwise Pearson correlation coefficients among the above inputs. As shown in the table, there are some significant correlations among specific usage fields. Statistically significant correlations (at a 0.01 level) are marked by an asterisk.

Table 3.1   Behavioral fields used in the PCA example.

Field name	Description
VOICE_OUT_CALLS	Monthly average of outgoing voice calls
VOICE_OUT_MINS	Monthly average number of minutes of outgoing voice calls
SMS_OUT_CALLS	Monthly average of outgoing SMS calls
MMS_OUT_CALLS	Monthly average of outgoing MMS calls
OUT_CALLS_ROAMING	Monthly average of outgoing roaming calls (calls made in a foreign country)
GPRS_TRAFFIC	Monthly average GPRS traffic
PRC_VOICE_OUT_CALLS	Percentage of outgoing voice calls: outgoing voice calls as a percentage of total outgoing calls
PRC_SMS_OUT_CALLS	Percentage of SMS calls
PRC_MMS_OUT_CALLS	Percentage of MMS calls
PRC_INTERNET_CALLS	Percentage of Internet calls
PRC_OUT_CALLS_ROAMING	Percentage of outgoing roaming calls: roaming calls as a percentage of total outgoing calls

---

Statistical Hypothesis Testing and Significance

Statistical hypothesis testing is applied when we want to make inferences about the whole population by using sample results. It involves the formulation of a null hypothesis that is tested against an opposite, alternative hypothesis. The null hypothesis states that an observed effect is simply due to chance or random variation of the particular dataset examined.

As an example of statistical testing, let us consider the case of the correlations between the phone usage fields presented in Table 3.2 and examine whether there is indeed a linear association between the number and the minutes of voice calls. The null hypothesis to be tested states that these two fields are not (linearly) associated in the population. This hypothesis is to be tested against an alternative hypothesis which states that these two fields are correlated in the population. Thus the statistical test examines the following statements:

H_0: the linear correlation in the population is 0 (no linear association); versus
H_a: the linear correlation in the population differs from 0.

The sample estimate of the population correlation coefficient is quite large (0.84) but this may be due to the particular data analyzed (one month of traffic data) and may not represent actual population relationships. Remember that the goal is to make general inferences and draw, with a certain degree of confidence, conclusions about the population. The good news is that statistics can help us with this.

By using statistics we can calculate how likely a sample correlation coefficient at least as large as the one observed would be, if the null hypothesis were to hold true. In other words, we can calculate the probability of such a large observed sample correlation if there is indeed no linear association in the population.

This probability is called the p-value or the observed significance level and it is tested against a predetermined threshold value called the significance level of the statistical test. If the p-value is small enough, typically less than 0.05 (5%), or in the case of large samples less than 0.01 (1%), the null hypothesis is rejected in favor of the alternative. The significance level of the test is symbolized by the letter α and it denotes the false positive probability (probability of falsely rejecting a true null hypothesis) that we are willing to tolerate. Although not displayed here, in our example the probability of obtaining such a large correlation coefficient by chance alone is small and less than 1%. Thus, we reject the null hypothesis of no linear association and consider the correlation between these two fields as statistically significant at the 0.01 level.

This logic is applied to various types of data (frequencies, means, other statistical measures) and types of problems (associations, mean comparisons): we formulate a null hypothesis of no effect and calculate the probability of obtaining such a large effect in the sample if indeed there was no effect in the population. If the probability (p-value) is small enough (typically less than 0.05) we reject the null hypothesis.

The number of outgoing voice calls for instance is positively correlated with the minutes of calls. The respective correlation coefficient is 0.84, denoting that customers who make a large number of voice calls also tend to talk a lot. Some other fields are negatively correlated, such as the percentage of voice and SMS calls (-0.98). This signifies a contrast between voice and SMS usage, not necessarily in terms of usage volume but in terms of the total usage ratio that each service accounts for. Conversely, other attributes do not seem to be related, like Internet and roaming calls for instance. Studying correlation tables in order to arrive at conclusions is a cumbersome job. That is where PCA comes in. It analyzes such tables and identifies groups of related fields.

PCA applied to the above data revealed five components by using the eigenvalue criterion, which we will present shortly. Table 3.3, referred to as the

Table 3.2 Pairwise correlation coefficients among the original input fields.

	VOICE OUT CALLS	VOICE OUT MINS	SMS OUT CALLS	OUT CALLS ROAMING	GPRS TRAFFIC	MMS OUT CALLS	PRC VOICE OUT CALLS	PRC SMS OUT CALLS	PRC OUT CALLS ROAMING	PRC INTERNET CALLS	PRC MMS OUT CALLS
VOICE OUT CALLS	1.00	0.84°	0.20°	0.16°	0.04°	0.14°	0.07°	−0.07°	−0.02°	−0.02°	−0.02°
VOICE OUT MINS	0.84°	1.00	0.22°	0.14°	0.05°	0.17°	0.00	−0.01	−0.01	−0.01	0.00
SMS OUT CALLS	0.20°	0.22°	1.00	0.18°	0.05°	0.28°	−0.72°	0.72°	0.03°	0.00	0.05°
MMS OUT CALLS	0.14°	0.17°	0.28°	0.03°	0.17°	1.00	−0.20°	0.19°	−0.01	0.06°	0.59°
OUT CALLS ROAMING	0.16°	0.14°	0.18°	1.00	0.00	0.03°	−0.04°	0.13°	0.66°	0.00	−0.01
GPRS TRAFFIC	0.04°	0.05°	0.05°	0.00	1.00	0.17°	−0.05°	0.05°	−0.01	0.53°	0.15°
PRC VOICE OUT CALLS	0.07°	0.00	−0.72°	−0.04°	−0.05°	−0.20°	1.00	−0.98°	0.02	−0.03°	−0.09°
PRC SMS OUT CALLS	−0.07°	−0.01	0.72°	0.13°	0.05°	0.19°	−0.98°	1.00	0.09°	0.03°	0.09°
PRC MMS OUT CALLS	−0.02°	0.00	0.05°	−0.01	0.15°	0.59°	−0.09°	0.09°	−0.02	0.17°	1.00
PRC INTERNET CALLS	−0.02°	−0.01	0.00	0.00	0.53°	0.06°	−0.03°	0.03°	0.00	1.00	0.17°
PRC OUT CALLS ROAMING	−0.02°	−0.01	0.03°	0.66°	−0.01	−0.01	0.02	0.09°	1.00	0.00	−0.02

Table 3.3 The component matrix.

	Component				
	1	**2**	**3**	**4**	**5**
PRC SMS OUT CALLS	0.89	−0.34	−0.17	−0.06	−0.10
PRC VOICE OUT CALLS	−0.88	0.36	0.11	0.15	0.11
SMS OUT CALLS	0.86	−0.01	−0.16	−0.16	−0.09
VOICE OUT CALLS	0.20	0.88	−0.04	−0.28	−0.12
VOICE OUT MINS	0.26	0.86	−0.02	−0.29	−0.11
GPRS TRAFFIC	0.19	0.09	0.60	0.35	−0.48
PRC INTERNET CALLS	0.12	0.02	0.58	0.40	−0.51
PRC OUT CALLS ROAMING	0.14	0.18	−0.44	0.77	0.11
OUT CALLS ROAMING	0.26	0.34	−0.46	0.66	0.08
PRC MMS OUT CALLS	0.28	0.04	0.59	0.19	0.60
MMS OUT CALLS	0.47	0.19	0.49	0.04	0.56

component matrix, presents the linear correlations between the original fields, in the rows, and the derived components, in the columns.

The correlations among the components and the original inputs are called loadings; they are typically used for the interpretation and labeling of the derived components. We will come back to loadings shortly, but for now let us examine why the algorithm suggested a five-component solution.

The proposed solution of five components is based on the eigenvalue criterion which is summarized in Table 3.4. This table presents the eigenvalues and the percentage of variance/information attributable to each component. The components are listed in the rows of the table. The highlighted first five rows of the table correspond to the extracted components. A total of 11 components are needed to fully account for the information of the 11 original fields. That is why the table contains 11 rows. However, not all these components are retained. The algorithm extracted five of them, based on the eigenvalue criterion which we specified when we set up the model.

The eigenvalue is a measure of the variance that each component accounts for. The eigenvalue criterion is perhaps the most widely used criterion for selecting which components to keep. It is based on the idea that a component should be considered insignificant if it does worse than a single field. Each single field contains one unit of standardized variance, thus components with eigenvalues below 1 are not extracted.

The second column of the table contains the eigenvalue of each component. Components are extracted in descending order of importance so the first one carries the largest part of the variance of the original fields. Extraction stops at component 5 since component 6 has an eigenvalue below the threshold of 1.

Table 3.4 The variance explained.

Total variance explained			
Components	**Eigenvalue**	**% of variance**	**Cumulative %**
1	**2.84**	**25.84**	**25.84**
2	**1.96**	**17.78**	**43.62**
3	**1.76**	**16.01**	**59.63**
4	**1.56**	**14.21**	**73.84**
5	**1.25**	**11.33**	**85.16**
6	0.49	4.45	89.62
7	0.38	3.41	93.03
8	0.34	3.06	96.09
9	0.26	2.38	98.47
10	0.16	1.44	99.92
11	0.01	0.08	100.00

Eigenvalues can also be expressed in terms of a percentage of the total variance of the original fields. The second column of the table denotes the proportion of the variance attributable to each component, and the next column denotes the proportion of the variance jointly explained by all components up to that point. The percentage of the initial variance attributable to the five extracted components is about 85%. This figure is not bad at all, if you consider that by keeping only 5 of the 11 original fields we lose just a small part of their initial information.

Technical Tips on the Eigenvalue Criterion

Variance is a measure of the variability of a field. It summarizes the dispersion of the field values around the mean. It is calculated by summing the squared deviations from the mean (and dividing them by the total number of records minus 1). Standard deviation is another measure of variability and is the square root of the variance. A standardized field with the z-score method is created with the following formula:

(Record value − mean value of field)/standard deviation of the field.

The variance can be considered as a measure of a field's information. A standardized field has a standard deviation and a variance value of 1, hence it carries one unit of information.

As mentioned above, each component is related to the original fields and these relationships are represented by the loadings in the component matrix.

The proportion of variance of a field that can be interpreted by another field is represented by the square of their correlation. The eigenvalue of each component is the sum of squared loadings (correlations) across all input fields. Thus, each eigenvalue denotes the total variance or total information interpreted by the respective component.

Since a single standardized field contains one unit of information, the total information of the original fields is equal to their number. The ratio of the eigenvalue to the total units of information (11 in our case) gives the percentage of variance that each component represents.

By comparing the eigenvalue to the value of 1 we examine if a component is more useful and informative than a single input.

Although the eigenvalue criterion is a good starting point for selecting the number of fields to extract, other criteria should also be evaluated before reaching the final decision. A list of commonly used criteria follows:

1. **The eigenvalue (or latent root) criterion:** This was discussed in the previous section. Typically the eigenvalue is compared to 1 and only components with eigenvalues higher than 1 are retained.

2. **The percentage of variance criterion:** According to this criterion, the number of components to be extracted is determined by the total explained percentage of variance. A successive number of components are extracted, until the total explained variance reaches a desired level. The threshold value for extraction depends on the specific situation, but, in general, a solution should not fall below 60–65%.

3. **The interpretability and business meaning of the components:** The derived factors should, above all, be directly interpretable, understandable, and useful. Since they will be used for subsequent modeling and reporting purposes, we should be able to recognize the information which they convey. A component should have a clear business meaning, otherwise it is of little value for further usage. In the next section we will present a way to interpret components and to recognize their meaning.

4. **The scree test criterion:** Eigenvalues decrease in descending order along with the order of the component extraction. According to the scree test criterion, we should look for a large drop, followed by a "plateau" in the eigenvalues, which indicates a transition from large to small values. At that point, the unique variance (variance attributable to a single field) that a component carries starts to dominate the common variance. This criterion is graphically illustrated by the scree plot which displays the eigenvalues against the number of extracted components. The scree plot for our example is presented in Figure 3.1. What

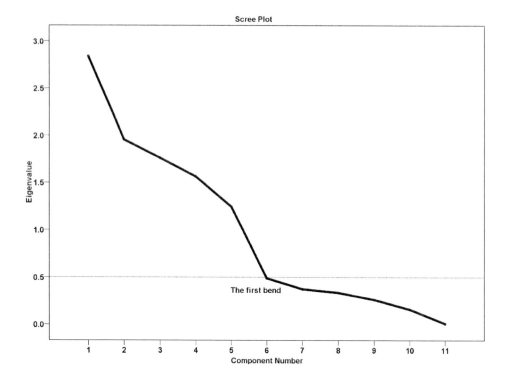

Figure 3.1   The scree plot for PCA.

we should look for is a steep downward slope in the eigenvalues' curve followed by a straight line. In other words, the first "bend" or "elbow" that would resemble the scree at the bottom of a mountain slope. The "bend" start point indicates the maximum number of components to extract while the point before the "bend" (in our example, five components) could be selected for a more "compact" solution.

In the case presented here, the scree test seems to support the eigenvalue criterion in suggesting a five-component solution. Additionally, the five components cumulatively account for 85% of the total variance, a value that is more than adequate. Moreover, a sixth component would complicate the solution and would only add a poor 4.5% of additionally explained variance.

In our example, all criteria seem to indicate a five-component solution. However, this is not always the case. The analyst should possibly experiment and try different extraction solutions before reaching a decision. Although an additional component might add a little complexity, it could be retained if it clarifies the solution, as opposed to a vague component which only makes things more confusing. In the final analysis, it is the transparency, the business meaning, and the usability that count. Analysts should be able to clearly recognize what each factor represents in order to use it in upcoming tasks.

WHAT IS THE MEANING OF EACH COMPONENT?

The next task is to determine the meaning of the derived components, with respect to the original fields. The goal is to understand the information that they convey and name them accordingly. This interpretation is based on the correlations among the derived components and the original inputs by examination of the corresponding loadings.

Rotation is a recommended technique to apply in order to facilitate the interpretation process of the components. Rotation minimizes the number of fields that are strongly correlated with many components and attempts to associate each input to one component. There are numerous rotation techniques, with Varimax being the most popular for data reduction purposes since it yields transparent components which are also uncorrelated, a characteristic usually required for subsequent tasks. Thus, instead of looking at the component matrix and its loadings, we will examine the rotated component matrix which results after application of a Varimax rotation.

Technical Tips on Rotation Methods

Rotation is a method used to simplify the interpretation of components. It attempts to clarify situations in which fields seem to be associated with more than one component. It tries to produce a solution in which each component has large $(+1, -1)$ correlations with a specific set of original inputs and negligible correlations (close to 0) with the rest of the fields.

As mentioned above, components are extracted in order of significance, with the first one accounting for as much of the input variance as possible and subsequent components accounting for residual variance. Subsequently, the first component is usually a general component with most of the inputs associated to it. Rotation tries to fix this and redistributes the explained variance in order to produce a more efficient and meaningful solution. It does so by rotating the reference axes of the components, as shown in Figure 3.2, so that the correlation of each component with the original fields is either minimized or maximized.

Varimax rotation moves the axes so that the angle between them remains perpendicular, resulting in uncorrelated components (orthogonal rotation).

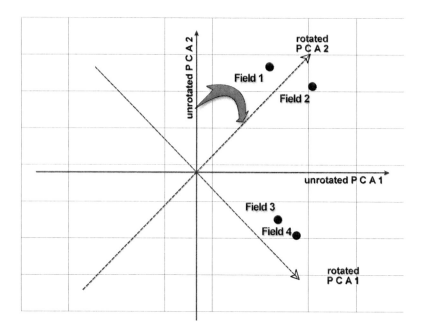

Figure 3.2 An orthogonal rotation of the derived components.

Other rotation methods (like Promax and Oblimin, for instance) are not constrained to produce uncorrelated components (oblique rotations) and are mostly used when the main objective is data interpretation instead of reduction.

Rotation reattributes the percentage of variance explained by each component in favor of the components extracted last, while the total variance jointly explained by the derived components remains unchanged.

In Table 3.5, loadings with absolute values below 0.4 have been suppressed for easier interpretation. Moreover, the original inputs have been sorted according to their loadings so that fields associated with the same component appear together as a set. To understand what each component represents we should identify the original fields with which it is associated, the magnitude, and the direction of the association. Hence, the interpretation process involves examination of the loading values and their signs and identification of significant correlations. Typically, correlations above 0.4 in absolute value are considered to be of practical significance and denote the original fields which are representative of each component. The interpretation process ends with the labeling of the derived components with names that appropriately summarize their meaning.

Table 3.5 The rotated component matrix.

	Rotated component matrix				
	Components				
	1	**2**	**3**	**4**	**5**
PRC VOICE OUT CALLS	−0.970				
PRC SMS OUT CALLS	0.968				
SMS OUT CALLS	0.850				
VOICE OUT CALLS		0.954			
VOICE OUT MINS		0.946			
PRC OUT CALLS ROAMING			0.918		
OUT CALLS ROAMING			0.900		
PRC MMS OUT CALLS				0.897	
MMS OUT CALLS				0.866	
PRC INTERNET CALLS					0.880
GPRS TRAFFIC					0.865

In this particular example, component 1 is strongly associated with SMS usage. Both the number (SMS OUT CALLS) and the ratio of SMS calls (PRC SMS OUT CALLS) load heavily on this component. Consequently, customers with high SMS usage will also have high positive values in component 1. The negative sign in the loading of the voice usage ratio (PRC VOICE OUT CALLS) indicates a strong negative correlation with component 1 and the aforementioned SMS fields. It suggests a contrast between voice and SMS calls, in terms of usage ratio: as SMS usage increases, the voice usage ratio tends to decrease. We can safely label this component as "SMS usage" as it seems to measure the intensity of use of that service.

Similarly, the number and minutes of voice calls (VOICE OUT CALLS and VOICE OUT MINS) seem to covary and are combined to form the second component, which can be labeled as "Voice usage." Component 3 measures "Roaming usage" and component 4 "MMS usage." Finally, the fifth component denotes "Internet usage" since it presents high positive correlations with both Internet calls (PRC INTERNET CALLS) and GPRS traffic (GPRS TRAFFIC).

Ideally, each field will load heavily on a single component and original inputs will be clearly separated into distinct groups. This is not always the case, though. Fields that do not load on any of the extracted components or fields not clearly separated are indications of a solution that fails to explain all the original fields. In these cases, the situation may be improved by requesting a richer solution with more extracted components.

Table 3.6   The communalities matrix.

	Communalities
VOICE OUT CALLS	0.913
VOICE OUT MINS	0.901
SMS OUT CALLS	0.791
OUT CALLS ROAMING	0.838
GPRS TRAFFIC	0.762
MMS OUT CALLS	0.808
PRC VOICE OUT CALLS	0.953
PRC SMS OUT CALLS	0.954
PRC OUT CALLS ROAMING	0.847
PRC INTERNET CALLS	0.778
PRC MMS OUT CALLS	0.823

DOES THE SOLUTION ACCOUNT FOR ALL THE ORIGINAL FIELDS?

The next question to be answered is whether the derived solution sufficiently accounts for all the original fields. This question is answered by Table 3.6, which contains the communalities of the original inputs.

Communalities represent the amount of variance of each field that is jointly accounted for by all the extracted components. High communality values indicate that the original field is sufficiently explained by the reduced PCA solution. Analysts should examine the communality table and assess whether the level of explained information is satisfactory (typically above 0.50) for each input.

Technical Tips on Communalities

Communality represents the total amount of variance of a specific field that is jointly accounted for by all the components. It is calculated as the sum of squared loadings of the field across all components.

A low communality value designates an original field with low loadings on the components and with an insignificant contribution to the formation of the PCA solution. Thus communality values can help in the identification of original fields left unexplained by the derived solution. Extraction of additional components may help in situations like these as it may account for the "forgotten" inputs. In the example presented here, all communalities are quite high so there should be no concern about the "representativeness" of the solution.

PROCEEDING TO THE NEXT STEPS WITH THE COMPONENT SCORES

The extracted components can be used in upcoming data mining models, provided of course that they comprise a conceptually clear and meaningful representation of the original fields. The PCA algorithm derives new composite fields, named component scores, that denote the values of each record in the revealed components. Component scores are produced through linear transformations of the original fields, by using coefficients that correspond to the loading values. They can be used as any other fields in subsequent tasks.

The derived component scores are continuous numeric fields with standardized values; hence they have a mean of 0 and a standard deviation of 1 and they designate the deviation from the average behavior. More specifically, the scores denote how many standard deviations above or below the overall mean each record lies. The list of the five component scores produced by PCA for 10 of the customers in our example is shown in Table 3.7.

The high score of customer 5 in component 1 denotes a customer with above-average SMS usage. The negative score in component 2 indicates low voice usage. Similarly, customer 2 seems to be a person who frequently uses their phone abroad (roaming usage measured by component 3) and customer 4 seems like a typical example of a "voice only" customer.

As noted above, the derived component scores, apart from being fewer in number than the original fields, are standardized, uncorrelated (due to Varimax

Table 3.7   A list of derived component scores.

Customer ID	Component score 1	Component score 2	Component score 3	Component score 4	Component score 5
1	0.633	−0.182	−0.263	1.346	−0.209
2	−0.964	−0.500	8.805	−0.090	−0.036
3	−0.501	−0.381	−0.196	−0.197	−0.063
4	−0.501	1.677	−0.272	−0.305	−0.055
5	3.660	−1.041	−0.385	−0.596	−0.084
6	−0.450	0.720	0.433	−0.251	−0.056
7	1.249	−0.276	1.043	−0.384	−0.028
8	−0.695	0.192	−0.204	0.461	−0.117
9	−0.902	−0.959	0.247	2.265	−0.164
10	0.028	0.212	2.715	1.186	−0.165

rotation), and equally represent all the underlying, input data dimensions. These characteristics make them perfect candidates for inclusion in subsequent clustering algorithms for segmentation.

RECOMMENDED PCA OPTIONS

Figures 3.3 and 3.4 and Table 3.8 present the recommended options for fine tuning the PCA model development process in IBM SPSS Modeler (formerly Clementine) and in any other data mining software which offers the specific technique. Although IBM SPSS Modeler integrates smart defaults appropriate for

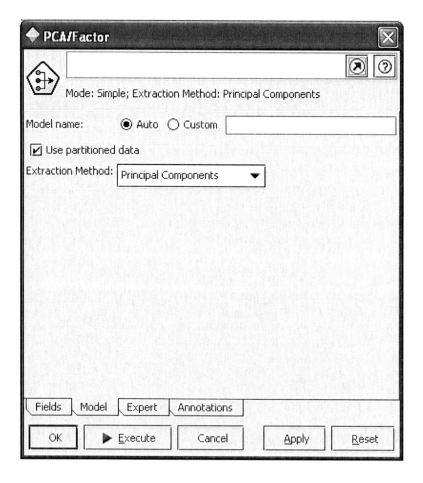

Figure 3.3   IBM SPSS Modeler recommended PCA/Factor Model options.

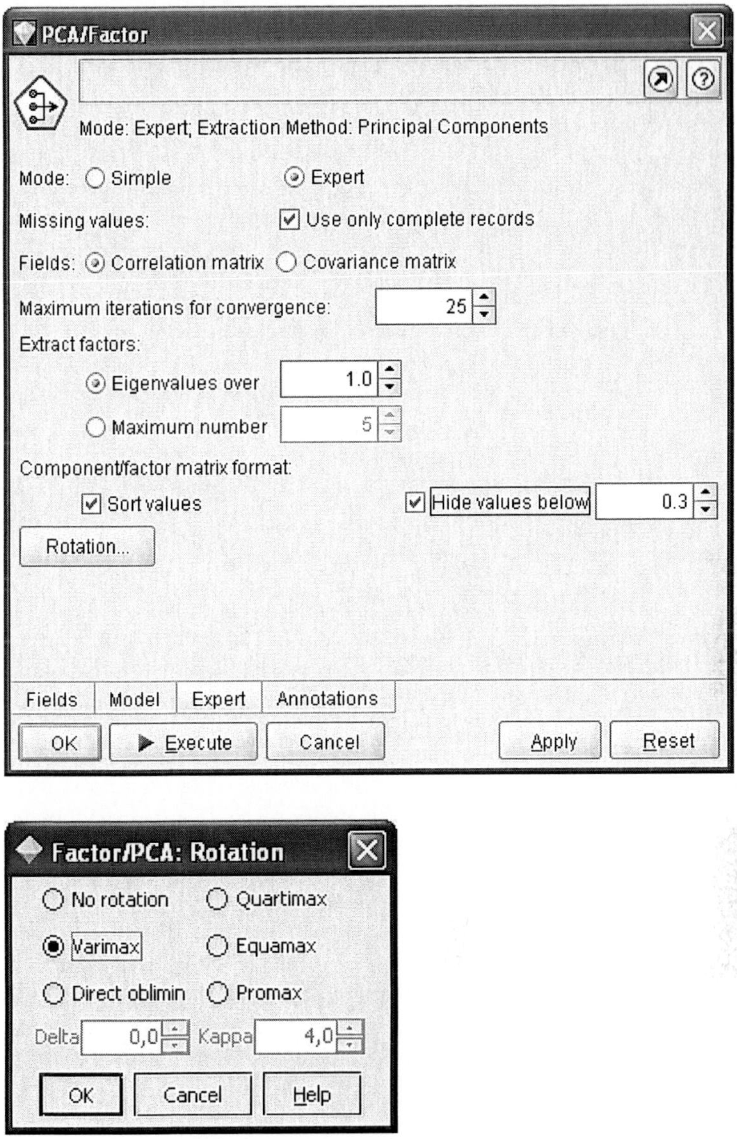

Figure 3.4   IBM SPSS Modeler recommended PCA Expert options.

most situations, specific modifications might help to produce improved results. These recommendations can only be considered as a suggested starting point and not as definitive guidelines suitable for any situation. Users should try various alternative approaches and test different parameter settings before selecting the final solution that best addresses their particular business goals.

Table 3.8   IBM SPSS Modeler recommended PCA options.

Option	Setting	Functionality/reasoning for selection
Extraction method	**PCA**	PCA is the recommended data reduction algorithm in cases where data reduction is the first priority *Worth trying alternatives: other data reduction algorithms like principal axis factoring and maximum likelihood are the most commonly used factor analysis methods*
Extract factors	**Eigenvalues over 1**	The most common criterion for determining the number of components to be extracted. Only components with eigenvalue above 1 are retained *Worth trying alternatives: users can intervene by setting a different threshold value. Alternatively, by examining alternative extraction criteria, they can specify a specific number of components to be retained, through the "Maximum number (of factors/components)" option*
Rotation	**Varimax**	Varimax rotation can facilitate the interpretation of the components by producing clearer loading patterns. It is an orthogonal rotation method which produces uncorrelated components
Sort values	**Selected**	Aids the interpretation of components by improving the readability of the rotated component matrix. It gathers together the original fields which are associated with the same component
Hide values below	**0.3**	Aids component interpretation by improving the readability of the rotated component matrix. It suppresses loadings with absolute values less than 0.3 and allows users to focus on significant loadings

CLUSTERING TECHNIQUES

In this section we will examine some of the most well-known clustering techniques. Specifically we will present the K-means, the TwoStep, and the Kohonen network algorithms. Moreover, since a vital part of a segmentation project is insight into the derived clusters and an understanding of their meaning, we will also propose ways for profiling the clusters and for outlining their differentiating characteristics.

DATA CONSIDERATIONS FOR CLUSTERING MODELS

Clustering models are unsupervised techniques, appropriate when there is no target output field. They analyze a set of inputs and group records with respect to the identified input data patterns.

The algorithms presented here (K-means, the TwoStep, Kohonen networks) work best with continuous numeric input fields. Although they can also handle categorical inputs, a general recommendation would be to avoid using categorical fields for clustering. The TwoStep model uses a specific distance measure that can more efficiently handle mixed (both categorical and continuous) inputs. K-means and Kohonen networks integrate an internal preprocessing encoding procedure for categorical fields. Each categorical field is recoded as a set of flag (binary/dichotomous) indicator fields, one such field for each original category. This recoding is called indicator coding. For each record, the indicator field corresponding to the category of the record is set to 1 and all other indicator fields are set to 0. Despite this special handling, categorical clustering fields tend to dominate the formation of the clusters and usually yield biased clustering solutions which overlook the differences attributable to the rest of the inputs.

Often, input clustering fields are measured in different scales. Since clustering models take into account the differences between records, differences in measurement scales can lead to biased clustering solutions simply because some fields might be measured in larger values. Fields measured in larger values show increased variability. If used in their original scale they will dominate the cluster solution. Thus, a standardization (normalization) process is necessary in order to put fields into comparable scales and ensure that fields with larger values do not determine the solution. The two most common standardization methods include the z-score and the 0–1 (or min–max) approaches. In the z-score approach, the standardized field is created as below:

(Record value − mean value of field)/standard deviation of the field.

Resulting fields have a mean of 0 and a standard deviation of 1. The record values denote the number of standard deviations above or below the overall mean value.

The 0–1 approach rescales all record values in the range 0–1, by subtracting the minimum from each value and dividing the difference by the range of the field:

(Record value − minimum value of field)/(maximum value of field
− minimum value of field).

IBM SPSS Modeler integrates internal standardization methods for all the algorithms presented here, so there is no need for the user to compensate for different scales. In general, though, standardization is a preprocessing step that should precede all clustering models. As a reminder, we note that principal component scores are standardized and this is an additional advantage gained by using PCA as the first step for clustering.

The presented algorithms also differ in regard to the handling of missing (null) values. In IBM SPSS Modeler, K-means and Kohonen networks impute the null values with "neutral" values. Missing values of numeric and categorical flag fields (dichotomous or binary) are substituted with a 0.5 value (remember that numeric fields are by default internally standardized in the range 0–1). For categorical set fields with more than two outcomes, the derived indicator fields are set to 0. Consequently, null values affect model training and, moreover, new cases with nulls are also scored and assigned to one of the identified clusters. Records with null values are not supported in TwoStep models. They are excluded from the TwoStep model training and new records with nulls are not scored and assigned to a cluster.

Another important issue to consider in clustering is the effect of possible outliers. Outliers are records with extreme values and unusual data patterns. They can be identified by examining data through simple descriptive statistics. IBM SPSS Modeler offers a data exploration tool (Data Audit) that provides basic statistics (average, minimum, and maximum value) and also identifies outlier values by examining deviations from the overall mean. Outliers can also be spotted by specialized modeling techniques like IBM SPSS Modeler's Anomaly Detection algorithm, which looks at full records and identifies unusual data patterns. Outlier records deserve special investigation as in many cases they are what we are looking for: exceptionally good customers or, at the other extreme, fraudulent cases. But they may have a negative impact on clustering models that aim at investigating "typical behaviors." They can confuse the clustering algorithm and lead to poor and distorted results. In many cases, the differences between "outlier" and "normal" data patterns are so large that they may mask the existing differences among the majority of "normal" cases. As a result, the algorithm can be guided to a degenerate solution that merely separates outliers from the rest of the records. Consequently, the clustering model may come up with a poor solution consisting of one large cluster of "normal" behavior and many very small clusters representing the unusual data patterns. This analysis may be useful, for instance, in fraud detection, but certainly it is not appropriate in the case of general purpose segmentation. Although the standardization process smoothes values and reduces the level of outlier influence in the formation of the clusters, a recommended approach for an enriched general purpose solution would be to identify outliers and treat them in a special manner. IBM SPSS Modeler's TwoStep algorithm integrates an internal outlier handling option for finding and specially treating outliers. Another approach would be to run an initial clustering solution, identify the small outlier clusters, and then rerun and fine-tune the analysis after excluding the outlier clusters.

CLUSTERING WITH K-MEANS

K-means is one of the most popular clustering algorithms. It starts with an initial cluster solution which is updated and adjusted until no further refinement is possible (or until the iterations exceed a specified number). Each iteration refines the solution by reducing the within-cluster variation.

The "K" in the algorithm's name comes from the fact that users should specify in advance the number of k clusters to be formed. The "means" part of the name refers to the fact that each cluster is represented by the means of its records on the clustering fields, a point referred to as the cluster central point or centroid or cluster center.

More specifically, unlike other clustering techniques (like TwoStep, for instance) which automatically determine the number of the underlying clusters through an automatic clustering procedure, K-means requires the user to specify in advance the number of clusters to be formed (k clusters). Therefore, this model requires a lot of trial and error and experimental modeling. Analysts should also use other automatic clustering models to get an indication of the number of clusters. Unavoidably, the quest for the optimal solution should involve preliminary analysis, multiple runs, and evaluation of different solutions.

K-means uses the Euclidean distance measure and iteratively assigns each record in the derived clusters. It is a very quick algorithm since it does not need to calculate the distances between all pairs of records. Clusters are refined through an iterative procedure during which records move between the clusters until the procedure becomes stable. The procedure starts by selecting k well-spaced initial records as cluster centers (initial seeds) and assigns each record to its "nearest" cluster. As new records are added to the clusters, the cluster centers are recalculated to reflect their new members. Then, cases are reassigned to the adjusted clusters. This iterative procedure is repeated until it converges and the migration of records between clusters no longer refines the solution.

The K-means procedure is graphically illustrated in Figure 3.5. For easier interpretation the graph refers to two input clustering fields and a two-dimensional clustering solution.

One drawback of this algorithm is that, to some extent, it depends on the initially selected cluster centers and the order of the input data. It is recommended that data are reordered randomly before model training. An even better approach

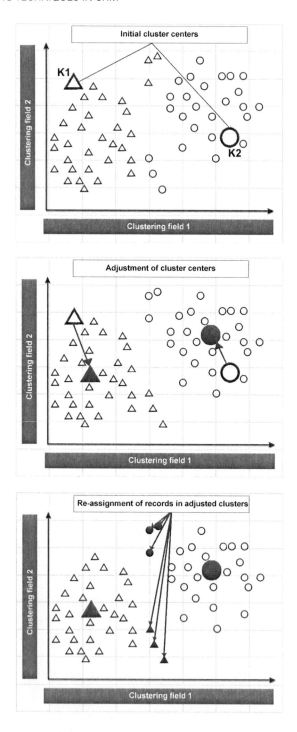

Figure 3.5   An illustration of the K-means clustering process.

would be to use another clustering model, save the resulting cluster centers, and load them as initial seeds for the K-means training.

On the other hand, K-means advantages include its speed and scalability: it is one of the fastest clustering models and it can efficiently handle long and wide datasets with many records and many input clustering fields.

Recommended K-means Options

Figures 3.6 and 3.7 and Table 3.9 present the recommended options for the IBM SPSS Modeler K-means modeling node. As outlined earlier, a key characteristic of

Table 3.9   IBM SPSS Modeler recommended K-means options.

Option	Setting	Functionality/reasoning for selection
Number of clusters	**Requires trial and error. Analysts should evaluate different clustering solutions**	Analysts should experiment with different numbers of clusters to find the solution that best fits their specific business goals *Worth trying alternatives: users can get an indication of the underlying number of clusters by trying other clustering techniques which incorporate specific criteria and automatically detect the number of clusters*
Generate distance field	**Selected**	It generates an additional field which denotes the distance of each record from the center of the assigned cluster. It can be used to assess whether a case is a typical representative of its cluster or lies far apart from the rest of the other members
Show cluster proximity	**Selected**	It creates a proximity matrix of the distances between the cluster centers. It can be used to assess the separation of the revealed clusters
Maximum iterations and change tolerance	**Keep the defaults, unless ...**	The algorithm stops if, after a data pass (iteration), the cluster centers do not change (0 change tolerance) or after the specified number of iterations has been completed, regardless of the change of the cluster centers *Users should browse the results and examine the solution steps and the change of the cluster centers at each iteration. In the case of non-convergence after the specified number of iterations, they should increase the specified number of iterations and rerun the algorithm*

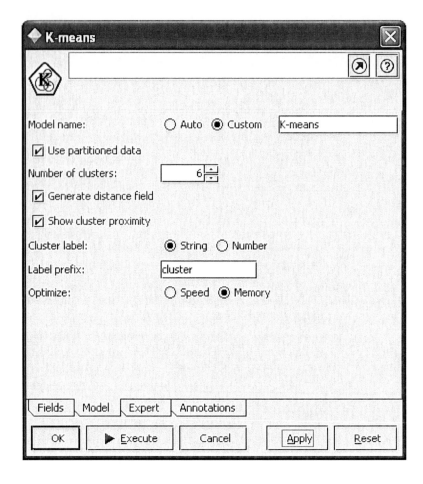

Figure 3.6   IBM SPSS Modeler recommended K-means Model options.

K-means is that it does not automatically provide an optimal clustering solution, thus it typically requires many runs before coming up with the most efficient grouping.

CLUSTERING WITH THE TWOSTEP ALGORITHM

The TwoStep algorithm involves a scalable clustering technique that can efficiently handle large datasets. It incorporates a special probabilistic distance measure that can efficiently accommodate both continuous and categorical input attributes.

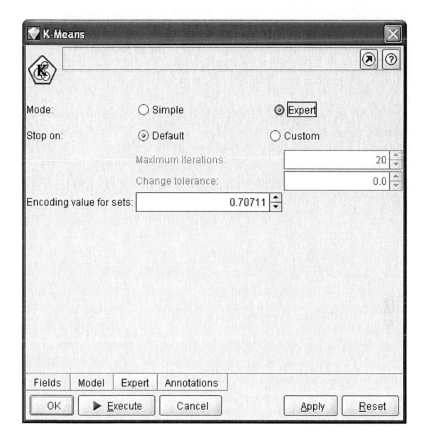

Figure 3.7 IBM SPSS Modeler recommended K-means Expert options.

Unlike classical hierarchical clustering techniques, it can handle a large amount of records due to the initial resizing of the input records to subclusters. As the name implies, the clustering process comprises two steps. In the first step of pre-clustering, the entire dataset is scanned and a large number of small, primary clusters are found. Records are assigned to the primary clusters based on their distance. The pre-clusters are characterized by respective summary statistics, namely the mean and variance of each numeric clustering field. Each record is recursively guided to the closest pre-cluster; if its distance is below an accepted threshold, it is assigned to the pre-cluster, otherwise it starts a pre-cluster of its own. A hierarchical (agglomerative) clustering algorithm is then applied to the pre-clusters, which are recursively merged until the final solution.

The TwoStep algorithm does not require the user to set in advance the number of clusters to fit. It can suggest a clustering solution automatically: the "optimal" number of clusters can be automatically determined by the algorithm according to a criterion that takes the following into account:

- The goodness of fit of the solution (the Bayes information criterion (BIC) or Schwartz Bayesian criterion), that is, how well the specific data are fitted by the current number of clusters.
- The distance measure for merging the subclusters. The algorithm examines the final steps in the hierarchical procedure and tries to spot a sudden increase in the merging distances. This point indicates that the agglomerative procedure has started to join dissimilar clusters and hence indicates the correct number of clusters to fit.

Another advantage of the TwoStep algorithm is that it integrates an outlier handling option that minimizes the effects of noisy records which otherwise could distort the segmentation solution. Outlier identification takes place in the pre-clustering phase. Small pre-clusters with few members compared to other pre-clusters (less than 25% of the largest pre-cluster) are considered as potential outliers. These outlier records are set aside and the pre-clustering procedure is rerun without them. Outliers that still cannot fit the revised pre-cluster solution are filtered out from the next step of hierarchical clustering and do not participate in the formation of the final clusters. Instead, they are assigned to a "noise" cluster. One drawback of this algorithm is that, to some extent, it depends on the order of the input data. It is recommended that data are reordered randomly before model training.

Recommended TwoStep Options

Figure 3.8 and Table 3.10 outline the recommended approach for applying the TwoStep clustering algorithm in IBM SPSS Modeler.

Table 3.10   IBM SPSS Modeler recommended TwoStep options.

Option	Setting	Functionality/reasoning for selection
Automatically calculate number of clusters	**Selected**	This option enables automatic clustering and lets the algorithm suggest the number of clusters to fit. The "maximum" and "minimum" text boxes allow the analysts to restrict the range of solutions to be evaluated. The default setting limits the algorithm to evaluate the last 15 steps of the hierarchical clustering procedure and to propose a solution comprising a minimum of 2 and up to a maximum of 15 clusters

Table 3.10 (*continued*)

Option	Setting	Functionality/reasoning for selection
		The automatic clustering option is the typical approach for starting model training
		Worth trying alternatives: apart from the automatic solution proposed by the algorithm, users can also evaluate alternative solutions by imposing a different range for automatic clustering. They can also "take the situation into their own hands," deactivate the automatic clustering, and explicitly ask for a specific number of clusters, by selecting the "Specify number of clusters" option and setting the required number of clusters to the respective text box
Exclude outliers	Selected	This option enables handling of outliers, an option generally recommended as it tends to yield richer clustering solutions

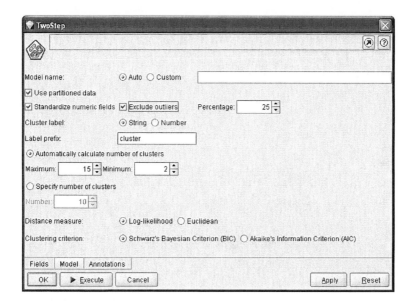

Figure 3.8   IBM SPSS Modeler recommended TwoStep Model options.

CLUSTERING WITH KOHONEN NETWORK/SELF-ORGANIZING MAP

Kohonen networks are special types of neural networks used for clustering. The network typically consists of two layers. The input layer includes all the clustering fields, called input neurons or units. The output layer is a two-dimensional grid map, consisting of the output neurons which will form the derived clusters. Each input neuron is connected to each of the output neurons with "strengths" or weights. These weights (which are analogous to the cluster centers referred to in the K-means procedure) are initially set at random and are refined as the model is trained.

Input records are presented to the output layer and the output neurons "compete" to "win" them. Each record "gravitates" toward the output neuron with the most similar pattern characteristics and is assigned to it. Record assignment is based on the Euclidean distance: each record's input values are compared to the centers of the output neurons and the "closest" output neuron "wins" the record.

This assignment also results in the adaptation or "learning" of the output neuron's characteristics and the adjustment of its corresponding weights (cluster center) so that they better match the pattern of the assigned record. Thus, the output neuron's center is updated and moved closer to the characteristics of the assigned record, in a way similar to that of K-means. Therefore, if another record with similar characteristics is presented to the output neuron, it will have a greater chance of winning it. The outline of a self-organizing map is graphically represented in Figure 3.9.

Data are passed through the network a specified number of times. The training process includes two phases, also referred to as cycles: a first cycle of rough estimations and large-scale changes in weights and a second refinement cycle with smaller changes.

After "winning" a record, the cluster center of the "winning" neuron is adapted by taking into account a portion of the difference between the current

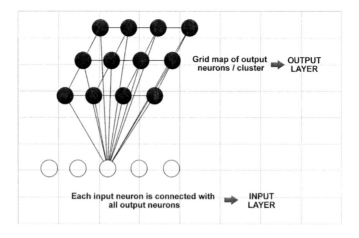

Figure 3.9   Self-organizing map graphical representation.

center and the record's input values. This portion, and thus the magnitude of the change in the weights, is determined by a change or learning rate parameter referred to as eta. Typically, the first phase has a relatively large eta to learn the overall data structure and the second phase incorporates a smaller eta to fine-tune the cluster centers.

Although quite similar, Kohonen networks and K-means also have significant differences. First of all, clusters in Kohonen networks are spatially arranged in a grid map. Moreover, the "winning" of records by a neuron/cluster also affects the weights of the surrounding neurons. Output neurons symmetrically around the "winning" neuron comprise a "neighborhood" of nearby units. Record assignment adjusts the weights of all neighboring neurons. Because of this neighborhood adaptation, the topology of the output map has a practical meaning, with similar clusters appearing close together as nearby neurons.

Output units with no winning records are removed from the solution. The retained output units represent the probable clusters. Users can specify the topology of the solution, that is, the maximum width and length dimensions of the output grid map. Selecting the right number of rows and columns for the output map requires trial and error.

Analysts should also evaluate the geometry/similarity and the density/ frequency of the proposed clusters. Kohonen networks involve many iterations and weight adjustments and consequently they are considerably slower than the TwoStep and K-means. Nevertheless, they are worth trying as a clustering alternative, especially because of the geometrical representation of the cluster similarity that they provide.

In Kohonen network models, cluster assignment is represented by two generated fields which denote the grid map co-ordinates (for instance, $X = 1$, $Y = 3$) of each record. These two fields should normally be concatenated into a single cluster membership field. A common and useful graphical representation of the geometry of the derived solution is through a simple scatterplot, with all records placed in the two-dimensional space defined by the grid co-ordinate fields. A scatterplot like that for a nine-cluster (3×3) solution is presented in Figure 3.10, depicting the values of the cluster membership field. This plot visually represents the density and the relative position, and hence similarity, of the resulting clusters.

Recommended Kohonen Network/SOM Options

Figures 3.11 and 3.12 and Table 3.11 explain the settings of the IBM SPSS Modeler Kohonen network/SOM model and provide suggestions for fine tuning of the algorithm.

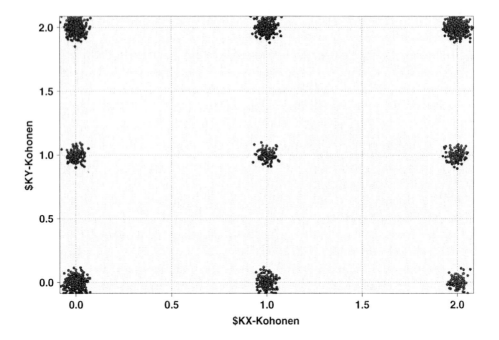

Figure 3.10   Sample scatterplot of a SOM grid map.

Table 3.11   IBM SPSS Modeler recommended Kohonen network options.

Option	Setting	Functionality/reasoning for selection
Set random seed	**Selected and a specific seed identified**	As mentioned before, the training of the Kohonen network model is initiated with random weights. This setting allows the analyst to set a specific value as a random seed for generating the initial weights, ensuring reproducible results on different runs
		If this option is not selected, the model's result will depend on the randomly selected seed, yielding, to a certain extent, different results on different runs, even if the other settings are held constant
Stop on	**Default**	Unless there is a strict time limitation, the model training process should not be stopped before concluding all the learning cycles and data passes. This option ensures that data are passed through the network an adequate number of times, ensuring model refinement

Table 3.11 (*continued*)

Option	Setting	Functionality/reasoning for selection
		The model training can also stop after a specified time has elapsed ("Stop on time (mins)" option). This option should only be selected if there is a specific time constraint
Width/length (of the output grid map)	**Requires trial and error**	Users can specify the size of the output grid map. This size determines the maximum number of probable clusters that will be formed by the model. Analysts can use prior knowledge from other clustering models to initially set the grid size In general it is recommended to start by asking for a quite rich solution, assuming that there are a significant number of member records in each cluster. The requested geometry of the solution, for instance 3×4 or 4×3, has less effect on the results than the requested grid size itself
Neighborhood/ initial eta/ cycles	**IBM SPSS Modeler default values**	Users can also intervene in the model training process by altering the following technical settings: neighborhood radius, initial eta value, and number of cycles. In general IBM SPSS Modeler default values are adequate for most situations and should not be modified. However, users could also try the following: • Neighborhood: defines the number of units symmetrically around a "winning" output neuron that will be adapted upon record assignment. Generally it should be an integer greater then 1. The neighborhood radius should be proportional to the requested grid size. So if we double the requested grid size we should also double the neighborhood. In small requested grids, the phase 1 neighborhood value could be set to 1 so that the entire grid is not affected • Cycles: determines the number of iterations or data passes through the network. Typically, both phase 1 and phase 2 default values (20 and 150 respectively) should not be modified since they are quite large and appropriate for most tasks

(continued overleaf)

Table 3.11   *(continued)*

Option	Setting	Functionality/reasoning for selection
		• The initial eta refers to the model's learning rate and the extent of change and adaptation of the weights to the characteristics of the "won" record. The model starts with coarse estimates and large weight changes which are gradually decreased to refine the solution. Thus, the learning rate at phase 1 starts with the initial eta value specified (0.3) and (linearly or exponentially) decreases, until completion of the specified phase 1 cycles, to the initial eta value specified for phase 2 (0.1). Similarly, the learning rate for phase 2 starts with the value specified and decreases to 0 (convergence) by the completion of phase 2 cycles
		Users can experiment by setting the initial eta values, for both phases, a bit higher, increasing the sensitivity of the output neurons and permitting larger adaptations to the presented data

EXAMINING AND EVALUATING THE CLUSTER SOLUTION

Analysts should examine and evaluate the revealed cluster solution and assess, among other things, the number and relative size of the clusters, their cohesion, and separation. A good clustering solution contains tightly cohesive and highly separated clusters. More specifically, the solution, through the use of descriptive statistics and specialized measures, should be examined in terms of the following.

THE NUMBER OF CLUSTERS AND THE SIZE OF EACH CLUSTER

Cluster examination normally begins by looking at the number of revealed clusters. In general purpose clustering, analysts typically expect a rich but manageable number of clusters. Analysts should look at the number of records assigned to each cluster. A large, dominating cluster which concentrates most of the records may indicate the need for further segmentation. Conversely, a small cluster with a few records merits special attention. If considered as an outlier cluster it could be set apart from the overall clustering solution and studied separately.

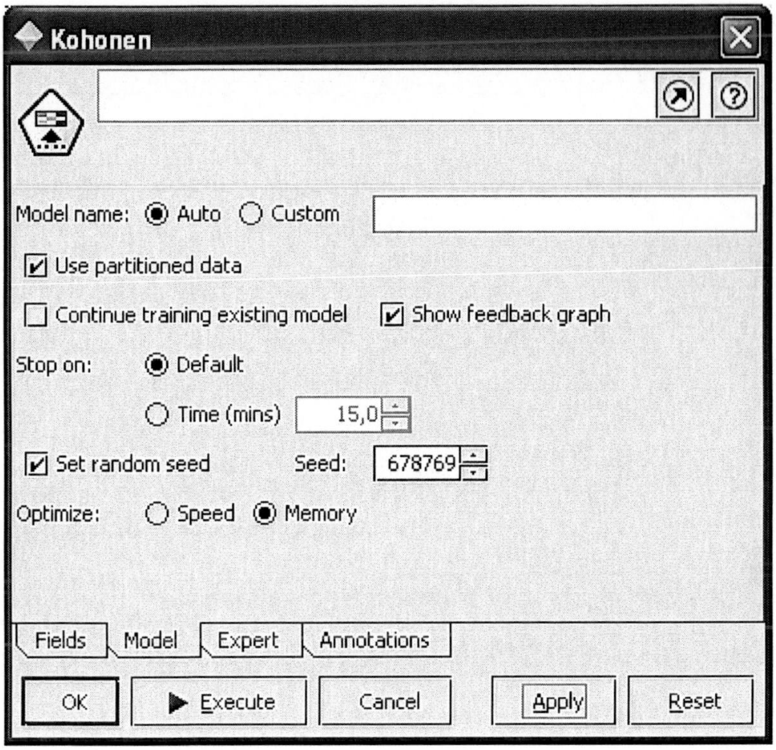

Figure 3.11   IBM SPSS Modeler recommended Kohonen Model options.

COHESION OF THE CLUSTERS

A good clustering solution is expected to be composed of dense concentrations of records around their centroids. Large dispersion values indicate non-homogeneous groupings and suggest further partitioning of the dataset. A number of useful statistics can be calculated to summarize the concentration and the level of internal cohesion of the revealed clusters such as:

• Standard deviations and pooled standard deviations of the clustering fields. Data miners should start by examining the standard deviations of the clustering fields for each cluster, hoping for small values which indicate a small degree of dispersion. The pooled standard deviation of a clustering field is the weighted (according to each cluster's size) average of the individual standard deviations for all clusters. Once again we anticipate low variability and small values which denote increased cohesion.

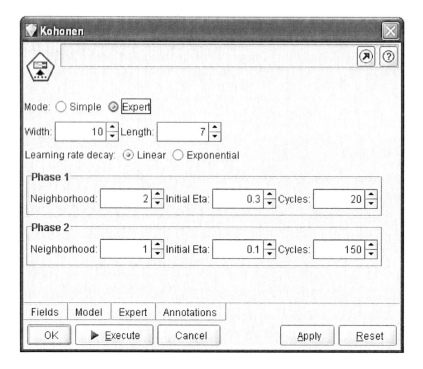

Figure 3.12   IBM SPSS Modeler recommended Kohonen Expert options.

- Maximum (Euclidean) distance from the cluster center. Another statistic that summarizes the degree of concentration of each cluster is the maximum distance from the cluster center, the cluster radius. In a way it represents the range of each cluster since it denotes how far apart the remotest member of the cluster lies.
- Evaluation of the distance between the members of a cluster and their cluster centroid. Analysts should average these distances over all members of a cluster and look for clusters with disproportionately large, average distances. These clusters are candidate for further segmentation. A technical cluster cohesion measure which is based on the (squared Euclidean) distances between the data points and their centroid is the sum of squares error (SSE). In order to compare models we can use the average SSE calculated as follows:

$$\text{Average SSE} = \frac{1}{N} \sum_{i \in C} \sum_{x \in C_i} dist(c_i, x)^2$$

where c_i is the centroid of cluster i, x a data point or record of cluster i, and N the total cases. A solution with smaller SSE is preferred.

SEPARATION OF THE CLUSTERS

Analysts also hope for well-separated (well-spaced) clusters:

- A good way of quantifying the cluster separation is to construct a proximity matrix with the distances between the cluster centroids. The minimum distance between clusters should be identified and assessed since this distance may indicate similar clusters that may be merged.
- Analysts may also examine a separation measure named sum of squares between (SSB), which is based on the (squared Euclidean) distances of each cluster's centroid to the overall centroid of the whole population. In order to compare models we can use the average SSB calculated as follows:

$$\text{Average SSB} = \frac{1}{N} \sum_{i \in C} N_i * dist(c_i, c)^2$$

where c_i is the centroid of cluster i, c the overall centroid, N the total cases, and N_i the number of cases in cluster i. The SSB is directly related to the pairwise distances between the centroids: the higher the SSB, the more separated the derived clusters.

A combined measure that assesses both the internal cohesion and the external separation of a clustering solution is the silhouette coefficient, which is calculated as follows:

1. For each record i in a cluster we calculate $a(i)$ as the average (Euclidean) distance to all other records in the same cluster. This value indicates how well a specific record fits a cluster. To simplify its computation, the $a(i)$ calculation may be modified to record the (Euclidean) distance of a record from its cluster centroid.
2. For each record i and for each cluster not containing i as a member, we calculate the average (Euclidean) distance of the record to all the members of the neighboring cluster. After doing this for all clusters where i is not a member, we calculate $b(i)$ as the minimum such distance in terms of all clusters. Once again, to ease computation, the $b(i)$ calculation can be modified to denote the minimum distance between a record and the centroid of every other cluster.
3. The silhouette coefficient for the record i is defined as:

$$\text{Si} = [b(i) - a(i)]/ \max\{a(i), b(i)\}$$

The silhouette coefficient varies between -1 and 1. Analysts hope for positive coefficient values, ideally close to 1, as this would indicate $a(i)$ values close to 0 and perfect internal homogeneity.

4. By averaging over the cases of a cluster we can calculate its average silhouette coefficient. The overall silhouette coefficient is a measure of the goodness of the clustering solution and can be calculated by taking the average over all records/data points. An average silhouette coefficient greater than 0.5 indicates reasonable partitioning, whereas a coefficient less than 0.2 denote a problematic solution.

Technical measures like the ones presented in this section are useful but analysts should also try to understand the true meaning of each cluster. An understanding of the clusters through profiling is required to take full advantage of them in subsequent marketing activities. Moreover, this is also a required evaluation step before accepting the solution. The process is presented in the next section.

Distance of Records from Their Cluster Center

A common approach for assessing the internal cohesion of a cluster is to examine the distance of its members from the cluster center. Ideally, member records should be concentrated around the cluster center, presenting low variability and standard deviation values.

Records "on the edge" of their cluster and far apart from their cluster center should be identified and examined separately. A specific measure and criterion could be employed to find these loose cluster members. A nice approach, similar to the one incorporated into IBM SPSS Modeler's Anomaly Detection model, considers as worth investigating records whose distance from their cluster centroid is at least n times (three or four times depending on the user's tolerance level) larger than the average distance of all member records.

A rather drastic but sometimes useful approach is to remove those outliers which lie on the boundary of their clusters. This may lead to an incomplete solution, which will not cover all records and leave some cases as "unclassified." Nevertheless, it will lead to more "solid" clusters with increased internal homogeneity and clear profiles.

UNDERSTANDING THE CLUSTERS THROUGH PROFILING

The last step in the development of the cluster solution is an understanding of the revealed clusters. This process demands the close co-operation of data miners and marketers in order to fully interpret the derived clusters and assess their business value, their usability, and actionability before deploying the derived segmentation scheme throughout the enterprise. It involves recognition of the differentiating characteristics of each cluster through profiling with the use of descriptive statistics and charts.

The cluster profiling typically involves:

1. **Comparison of clusters with respect to the clustering fields – examining the cluster centers:** The profiling phase usually starts with the cross-tabulation of the cluster membership field with the clustering inputs which are also referred to as the clustering fields. The goal is to identify the input data patterns that distinguish each cluster. Derived clusters will be interpreted and labeled according to their differentiating characteristics and consequently the profiling phase inevitably includes going back to the inputs and determining the uniqueness of each cluster with respect to the clustering fields.

    Typically, data miners start the profiling process with an examination of the cluster centers or seeds, also referred to as the cluster centroids. A cluster centroid is defined by simply taking the input fields and averaging over all the records of a cluster. The centroid can be thought of as the prototype or the most typical representative member of a cluster. Cluster centers for the simple case of two clustering fields and two derived clusters are depicted in Figure 3.13.

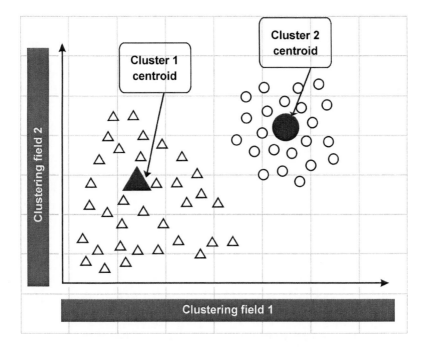

Figure 3.13   Graphical representation of the cluster centers or centroids.

Table 3.12 The table of cluster centers.

	Cluster 1	Cluster 2 ...	Cluster N	Overall population – all clusters
Clustering field 1	Mean(1, 1)	Mean(1, 2)	Mean(1, N)	**Mean(1)**
Clustering field 2	Mean(2, 1)	Mean(2, 2)	Mean(2, N)	**Mean(2)**
⋮	⋮	⋮	⋮	⋮
Clustering field M	Mean(M, 1)	Mean(M, 2)	Mean(M, N)	**Mean(M)**

In the general case of M clustering fields and N derived clusters, a table summarizing the cluster centers has the form shown in Table 3.12.

Analysts should check each cluster individually and compare its means for the input attributes to the overall population means, looking for significant deviations from the "typical" behavior. Therefore, they should search for clustering fields with relatively low or high mean values in a cluster.

2. **Comparison of clusters with respect to other key performance indicators:** The profiling phase should also examine the clusters with respect to "external" fields not directly participating in cluster formation, including key performance indicators (KPIs) and demographic information of interest. Normally, the cluster separation is not only limited to the clustering fields, but also reflected in other attributes.

Therefore, data miners should also describe the clusters by using all the important attributes, regardless of their involvement in the cluster building, to fully portray the structure of each cluster and identify the features that best characterize them.

This profiling typically involves examination of the mean of each continuous field of interest for each cluster. For categorical attributes, the procedure is analogous, involving comparisons of frequencies and percentages. The scope again is to uncover the cluster differentiation in terms of categorical fields.

This thorough examination of the clusters concludes with their labeling. All clusters are assigned names that adequately summarize their distinctive characteristics in a concise and simple way that will facilitate their subsequent use.

PROFILING THE CLUSTERS WITH IBM SPSS MODELER'S CLUSTER VIEWER

IBM SPSS Modeler offers a cluster profiling and evaluation tool that graphically presents the structure of the revealed clusters. This tool, named the Cluster Viewer,

Model Summary

Algorithm	K-Means
Input Features	5
Clusters	6

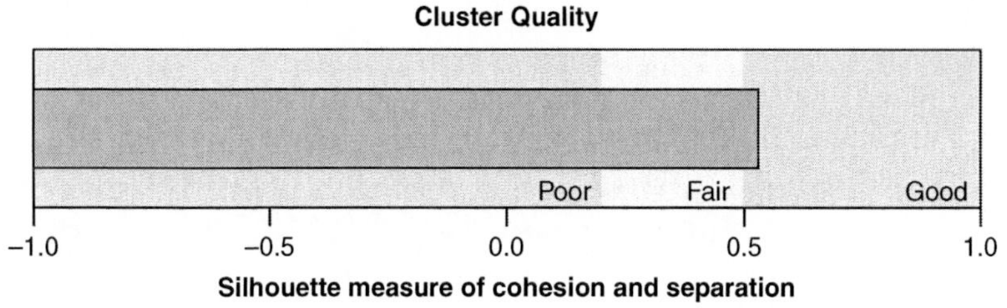

Cluster Quality

Silhouette measure of cohesion and separation

Figure 3.14   The overall silhouette measure of the clustering solution.

provides a useful visual examination of the clusters. The following figures show the Cluster Viewer results for the mobile phone case study introduced in "How Many Components Are to Be Extracted?".

Six clusters were derived after applying a K-means clustering algorithm to the principal component scores. The cluster quality of the solution (Figure 3.14) is good with an overall silhouette coefficient slightly greater than 0.5.

The next output of the Cluster Viewer is a cluster-by-inputs grid (Figure 3.15) that summarizes the cluster centroids. The rows contain all the clustering features/input fields, the five principal components, and the columns contain the revealed clusters. The first row of the table illustrates the size of each cluster. The table cells denote the cluster centers: the mean values of the component scores over all members of each cluster.

At first sight the differentiating factors of each cluster are evident. Cluster 1 includes a relatively small number of "superactive" customers with increased usage of all services. Cluster 2 customers are predominantly voice users. On the contrary, cluster 3 mainly includes SMS users. Cluster 4 customers show increased roaming usage while cluster 5 customers show increased usage of MMS and Internet services. The customers in cluster 6 present the lowest usage of all services.

Cluster interpretation can be aided by additional Cluster Viewer graphs such as the ones presented below. The graph in Figure 3.16 for instance compares clusters 2 and 3 for the first two components, SMS and voice usage.

Clusters

Feature
Importance
☐ 1

cluster	cluster-1	cluster-2	cluster-3	cluster-4	cluster-5	cluster-6
Label	Active users	Voice users	SMSers	Roamers	Tech users	Basic users
Size	1.6% (250)	23.5% (3660)	24.5% (3817)	7.6% (1179)	8.2% (1273)	34.7% (5405)
Features	Component 1-SMS usage 0.52	Component 1-SMS usage −0.15	Component 1-SMS usage 1.83	Component 1-SMS usage 0.03	Component 1-SMS usage 0.24	Component 1-SMS usage −0.43
	Component 2-Voice usage 0.77	Component 2-Voice usage 1.71	Component 2-Voice usage −0.31	Component 2-Voice usage −0.20	Component 2-Voice usage −0.07	Component 2-Voice usage −0.24
	Component 3-Roaming usage 5.58	Component 3-Roaming usage −0.20	Component 3-Roaming usage −0.25	Component 3-Roaming usage 2.48	Component 3-Roaming usage −0.13	Component 3-Roaming usage −0.16
	Component 4-MMS usage 3.25	Component 4-MMS usage −0.11	Component 4- MMS usage −0.26	Component 4-MMS usage −0.13	Component 4-MMS usage 2.44	Component 4-MMS usage −0.12
	Component 5-Internet usage 2.34	Component 5-Internet usage 0.00	Component 5-Internet usage −0.03	Component 5-Internet usage −0.03	Component 5-Internet usage 0.20	Component 5-Internet usage −0.03

Figure 3.15   IBM SPSS Modeler's representation of the cluster centers.

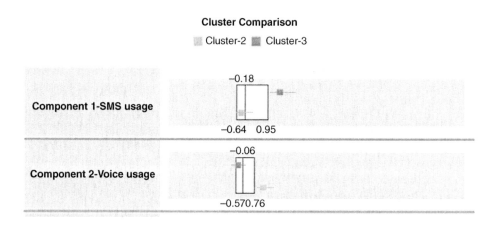

Cluster Comparison

▨ Cluster-2   ▨ Cluster-3

Component 1-SMS usage

−0.18

−0.64   0.95

Component 2-Voice usage

−0.06

−0.570.76

Figure 3.16   Cluster comparison with boxplots.

The background plot is a boxplot that summarizes the entire population. The vertical line inside the box represents the population median on the respective clustering field. The median is the 50th percentile, the value which separates the population into two sets of equal size. The width of the box is equal to the interquartile range, the difference between the 25th and the 75th percentiles, and indicates the degree of dispersion in the data. Thus the box represents the range including the middle 50% of the entire population. Overlaid on the background boxplots are the boxplots for the selected clusters. Point markers indicate the corresponding medians and the horizontal spans the interquartile ranges. As shown in the figure, cluster 3 is characterized by relatively high values on principal component 1 and lower values on component 2. Thus, if we recall the interpretation of the components presented earlier, cluster 3 mostly includes heavy SMS users with low voice traffic. On the contrary, cluster 2 presents relatively higher values on factor 2 and hence increased voice usage.

Another useful Cluster Viewer graph is shown in Figure 3.17. It examines the distribution of the component 3 scores (roaming usage) for cluster 4. The darker curve represents the cluster 4 distribution and is overlaid on the curve of the entire customer base. The cluster 4 curve is on the right tail of the overall population curve designating increased roaming usage for those customers.

ADDITIONAL PROFILING SUGGESTIONS

Even if the clustering solution has been built on principal component scores/factors, it is always a good idea to go back to the original inputs and summarize the clusters in terms of the original data. Usually, the component scores can give an overview and a useful first insight into the meaning of the clusters; however, examining clusters with respect to the original attributes can provide a more direct interpretation that can be more easily communicated. Original fields can be standardized before profiling for easier and more effective comparison, especially if they are measured on different scales and have different variability.

Table 3.13 refers to the six clusters of our simple telecommunications case study and summarizes the averages of important original attributes over the records of each cluster. The results of this table reinforce the ones indicated by the Cluster Viewer graphs.

The profiling process can also be enriched by and facilitated with charts like those suggested in the following sections. For instance, a nice visual exploration of the cluster solution, applicable to continuous profiling attributes, is provided by plotting the percentage deviation of each cluster from the overall mean values. The respective set of plots for the six clusters of the telecommunications example is presented in Figure 3.18.

Another useful profiling set of charts and statistical tests are offered by the SPSS AIM command. The mean values of the selected continuous profiling attributes are compared to the overall population values (based on *t*-tests) and statistically significant differences are detected, enabling analysts to focus on the most differentiating fields. A sample output from the SPSS AIM command is shown in Figure 3.19, presenting the results for cluster 3 in our example. The

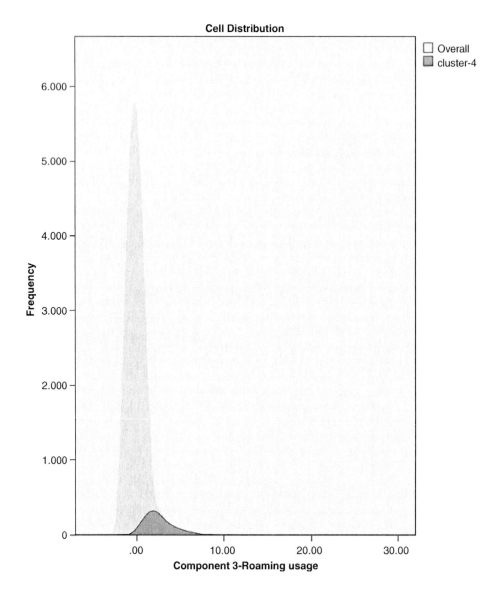

Figure 3.17   Examining the distribution of a clustering field for a specific cluster.

vertical lines represent the statistical test's critical values and the height of each bar the value of the respective test statistic (Student's t). The bar on the right panel of the chart surpassing the vertical line indicates a mean value in the cluster that is statistically different (higher) from the overall mean. On the other side, a bar on the left panel of the chart surpassing the vertical line indicates a mean value in the cluster lower than the overall population mean.

Table 3.13   The means of original attributes of particular importance for each cluster.

	Clusters						
	Cluster 1	Cluster 2	Cluster 3	Cluster 4	Cluster 5	Cluster 6	Total
VOICE CALLS	345.3	329.1	105.5	133.8	136.7	107.4	**165.6**
SMS CALLS	292.9	44.9	128.1	42.3	57.8	7.8	**55.4**
VOICE ROAM- ING CALLS	104.5	1.3	0.8	25.7	1.2	0.4	**3.5**
MMS CALLS	8.4	0.5	0.3	0.2	4.1	0.1	**0.7**
GPRS TRAFFIC	807.0	55.7	34.9	21.3	310.4	16.9	**62.4**

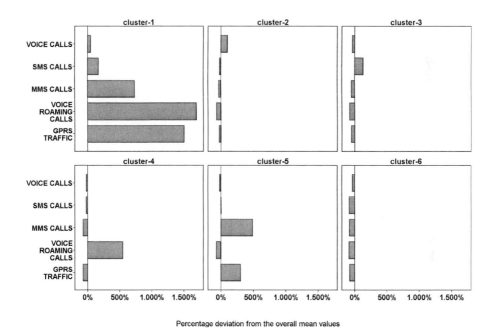

Percentage deviation from the overall mean values

Figure 3.18   Percentage deviation of each cluster from the overall mean values.

Simple scatterplots which place the cases of each cluster within the axes defined by selected attributes are also typically employed for visualization of the clusters' structure. The next set of scatterplots in Figure 3.20 graphically illustrates the relationship of the clusters in our example with voice and SMS usage. Due to the large size of the customer population, the scatterplots are based on a random sample of the assigned records. Moreover, the values of the selected attributes are

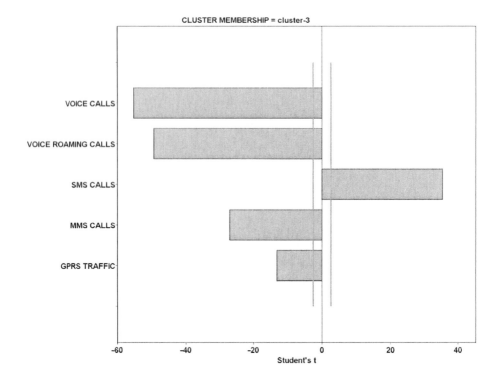

Figure 3.19   Comparing the cluster's mean values to the overall mean values through the SPSS AIM command.

standardized (z-scores) in order to reduce the measurement scale differences and more clearly portray the structure of the clusters.

All the above sets of charts, although not discussed in detail, reinforce and further clarify the initial cluster interpretation. Based on the profiling information, it is up to the project team involved to decide on the effectiveness of the solution. If the solution makes sense and covers the business objectives it will be adopted and deployed in the organization to support the marketing strategies. If not, the segmentation process should be reviewed to search for a more effective segmentation scheme.

SELECTING THE OPTIMAL CLUSTER SOLUTION

Regardless of the clustering method, data miners should also assess the homogeneity/cohesion of the clusters and the level of similarity of their members, as well as their separation, by examining the measures presented in the relevant section.

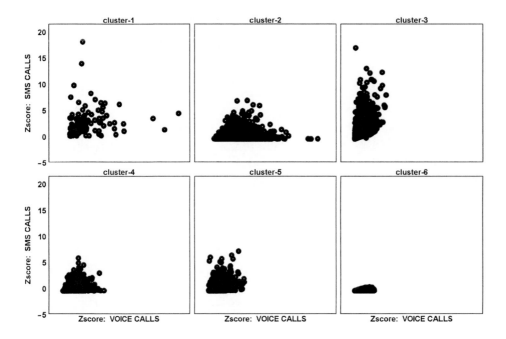

Figure 3.20   Using a scatterplot to examine the structure of the clusters.

Although all the above evaluation measures make mathematical sense and provide technical help to identify the optimal solution, data miners should not solely base their decision on them. A clustering solution is justified only if it makes sense from a business point of view. Actionability, potential business value, interpretability, and ease of use are factors that are hard to quantify and, in a way, measure subjectively. However, they are the best benchmarks for determining the optimal clustering solution. The profiling of the clusters and identification of their defining characteristics are essential parts of the clustering procedure. It should not be considered as a post-analysis task but rather as an essential step for assessing the effectiveness of the solution; that is why we have dedicated a whole section of this chapter to these topics.

As with every data mining model, data miners should try many different clustering techniques and compare the similarity of the derived solutions before deciding which one to choose. Different techniques that generate analogous results are a good sign for identifying a general and valid solution. As in any other data mining model, clustering results should also be validated by applying the model in a disjoint dataset and by examining the consistency of the results.

CLUSTER PROFILING AND SCORING WITH SUPERVISED MODELS

Apart from tables and charts of descriptive statistics, the cluster profiling process could also involve appropriate supervised models. Data miners can build classification models such as decision trees, for instance, with the cluster membership field as the target and all fields of interest as inputs, to gain more insight into the revealed clusters. These models can jointly assess many attributes and reveal those which best characterize each cluster. In the next section we will present a brief introduction to decision trees. Because of their transparency and the intuitive form of their results, decision trees are commonly applied for an understanding of the structure of the revealed clusters.

Additionally, decision trees can also be used as a scoring model for allocating new records to established clusters. Decision trees can translate the differentiating characteristics of each cluster into a set of simple and understandable rules which can subsequently be applied for classifying new records in the revealed clusters. Although this approach also introduces a source of errors due to possible weaknesses in the derived decision tree model, it is also a more transparent approach for cluster updating. It is based on understandable, model-driven rules, similar to common business rules, which can more easily be examined and communicated. Additionally, business users can more easily intervene and, if required, modify these rules and fine tune them according to their business expertise.

For all the above reasons, the next sections are dedicated to decision trees since they make up an excellent supplement to cluster analysis.

AN INTRODUCTION TO DECISION TREE MODELS

Decision trees belong to the class of supervised algorithms and are one of the most popular classification techniques. A decision tree consists of a set of rules, expressed in plain English, that associate a set of conditions to a specific outcome. These rules can also be represented in an intuitive tree format, enabling the visualization of the relationships between the predictors and the output.

Decision trees are often used for insight and for the profiling of target events/ attributes due to their transparency and the explanation of the predictions that they provide. They perform a kind of "supervised segmentation": they recursively partition (split) the overall population into "pure" subgroups, that is, homogeneous subgroups with respect to the target field, for instance subgroups containing only churners or only members of a particular cluster.

They start with an initial or root node which contains the entire dataset. At each level, the effect of all predictors on the target field is evaluated and the population is split with the predictor that yields the best separation in regard to the target field. The process is applied recursively and the derived groups are split further into smaller and "purer" subgroups until the tree grows fully into a set of "terminal" nodes. In general, the tree growing procedure stops when additional splits can no longer improve the separation or when one of the user-specified stopping criteria has been met.

Decision trees, when used in classification modeling, generate a rule set containing rules in the following format:

```
IF (PREDICTOR VALUES) THEN (PREDICTION=TARGET OUTCOME with a specific
CONFIDENCE or PROPENSITY SCORE).
```

For example:

```
IF (Tenure <= 4 yrs and Number of different products <= 2 and Recency of
last Transaction > 20 days) THEN (PREDICTION=Churner and CONFIDENCE=0.78).
```

When the tree is fully grown, each record lands in one of the terminal nodes. Each terminal node comprises a distinct rule which associates the predictors to the output. All records landing in the same terminal node get the same prediction and the same prediction confidence (likelihood of prediction).

The path of successive splits from the root node to each terminal node indicates the rule conditions; in other words, the combination of predictor characteristics associated with a specific outcome. In general, if misclassification costs have not been defined, the modal, the most frequent, outcome category of the terminal node denotes the prediction of the respective rule. The proportion of the modal category in each terminal node designates the confidence of the prediction. Thus, if 78% of records in a terminal node fall within the category of non-churners, then the respective rule's prediction would be "no churn" with a confidence of 78% or 0.78. Since the target field in this example is binary and simply separates churners from non-churners, the rule's churn propensity would be (1–0.78), thus 0.22 or 22%. When the model scores new cases, all customers which satisfy the rule conditions, that is all customers with the terminal node's profile, will be assigned a churn propensity of 22%.

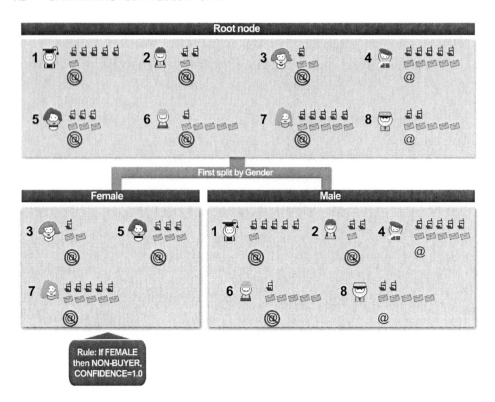

Figure 3.21   First split of the root node in a decision tree model.

To illustrate the tree growth procedure we will revisit the simple cross-selling example presented in "Predicting Events with Classification Modeling" in Chapter 2. The goal is to use the responses collected from a pilot campaign carried out on a sample of customers to identify the right prospects for promoting the Internet service. The first step of the decision tree model is presented in Figure 3.21.

Initially, all available inputs are assessed with respect to their predictive ability and gender is selected as the first split according to the algorithm's criteria. The entire dataset is partitioned into two subsets (child nodes) according to gender. Since all women rejected the offer, the "women" node is terminal due to perfect separation and cannot be partitioned any further. Hence the first rule has been discovered and classifies all women as non-buyers with a respective prediction confidence of 1.0 and a purchase propensity of 0. On the other side of the tree, three out of the five men (60%) rejected the offer, thus if the model had stopped at this level, the prediction would have been no purchase with a respective confidence of $3/5 = 0.6$. However, there is room for improvement (we also suppose that none of our stopping criteria have been met) and the algorithm proceeds by further

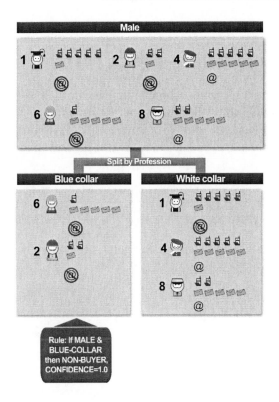

Figure 3.22   Further growing of the decision tree by partitioning a parent node.

splitting the "men" branch of the tree into smaller subgroups and child nodes (Figure 3.22).

At this level of the tree, occupation category is the best predictor since it results in optimal separation. The parent branch is partitioned into two child nodes: blue-collar and white-collar men. The node of blue-collar men presents absolute cohesion since it only contains (100%) non-buyers. Thus one more rule has been identified, which classifies all blue-collar men as non-buyers with a confidence of 1.0. The percentage of buyers rises at about 67% (2/3) among white-collar men who seem to constitute the target group of the promoted service. In our fictional example the purity of this node can be further improved and this last partitioning is presented in Figure 3.23.

Finally, white-collar men are segmented according to their SMS usage, resulting in two completely homogeneous terminal nodels. The percentage of buyers reaches 100% among white-collar men with high SMS usage, inducing a confident rule for identification of good Internet service prospects.

The above oversimplified example may be useful in clarifying the way that decision trees work, yet it cannot be considered as realistic. On the contrary we

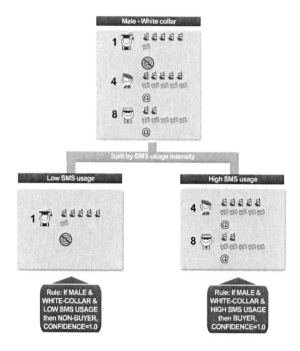

Figure 3.23   Final splits of the decision tree and the last terminal nodes.

deliberately kept it as simple as possible by presenting only faultless rules and by eliminating uncertainties due to model errors and misclassifications. In real projects we would not dare to split nodes with just a handful of records. Moreover, the resulting terminal nodes would rarely be so homogeneous. Terminal nodes typically include records of all the outcome categories but with increased purity and decreased diversity in the distribution of the categories of the outcome, compared to the overall population. In these situations the majority rule (modal category) defines the prediction of each terminal node. In real-word applications, models make errors. They misclassify cases and inevitably produce confidence values lower than 1.0. That is why data miners should thoroughly evaluate the performance of the resulting models and assess their predictive ability before deployment.

Figure 3.24 summarizes the entire tree growth procedure of our example, from the root node up to the terminal nodes.

The following figures show what the decision tree model results look like in real life. The first figure (Figure 3.25) presents the model results in the Tree format while the second one (Figure 3.26) presents the results in the equivalent

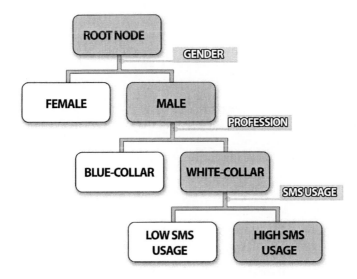

Figure 3.24   An outline of the resulting decision tree model.

Ruleset format. Eventually, the training dataset (root node) is partitioned into four subgroups (terminal nodes), resulting in four corresponding rules, three of which classify customers as non-prospective buyers and one as prospective buyers.

In general, the split-and-grow procedure of a node terminates when:

- All records in a specific node have the same value for the target field, for instance the node contains only churners, indicating perfect separation and absolute purity of the node.
- A significant predictor cannot be found and the separation, although not perfect, cannot be improved further.
- One of the specific user-defined stopping criteria has been met. Stopping criteria are used in order to specify:

  - The number of allowed successive splits: users can specify in advance the maximum tree depth (tree levels below the root) which determines the maximum number of times the training dataset can be recursively split.
  - The number of records of the nodes: users can specify in advance the minimum number of records of the parent (the nodes to be split) and of the child (the nodes resulting after partition) nodes.

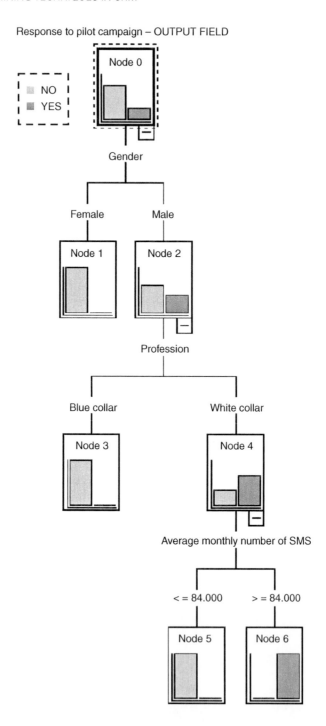

Figure 3.25 A decision tree model in IBM SPSS Modeler (Tree format).

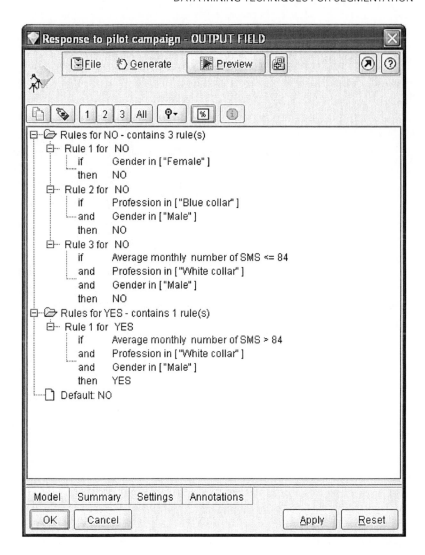

Figure 3.26   A decision tree model in IBM SPSS Modeler (Ruleset format).

Handling of Predictors by Decision Tree Models

Decision tree models can handle both categorical and numeric continuous predictors.

The categories of each categorical predictor are assessed with respect to the output and the homogeneous groups are merged (regrouped) before evaluation of the specific categorical predictor for splitting. As an example, let us consider a hypothetical churn prediction model which assesses the effect of marital status on churn. The marital status field includes four different categories: single, married, divorced, and widowed. The algorithm, based on specific criteria, concludes that only single customers present a different behavior with respect to churn. Thus, it regroups the marital status and if this field is selected for splitting, it will provide a binary split which will separate single customers from the rest. This regrouping process simplifies the understanding of the generated model and allows analysts to focus on true discrepancies among groups that really differ with respect to the output.

Decision tree models discretize continuous predictors by collapsing their values into ordered categories before evaluating them for possible splitting. Hence the respective fields are transformed to ordinal categorical fields, that is, fields with ordered categories. As an example let us review the handling of the continuous field which represents the number of SMS messages in the telecommunications cross-selling exercise presented above. A threshold of 84 SMS messages was identified and the respective split partitioned customers into two groups: those with more and those with less than 84 SMS messages per month.

Developing Stable and Understandable Decision Tree Models

In the case of decision tree models, "less is more" and the simplicity of the generated rules is also a factor to consider besides predictive ability. The number of tree levels should rarely be set above five or six. In cases where decision trees are mainly applied for profiling and an explanation of a particular outcome, this setting should be kept even lower (by requesting three levels for instance) in order to provide a concise and readable rule set that will enlighten the associations between the target and the inputs.

A crucial aspect to consider in the development of decision tree models is their stability and their ability to capture general patterns, and not patterns pertaining only to the particular training dataset. In general, the impurity decreases as the tree size increases. Decision trees, if not restricted with appropriate settings, may continue to grow until they reach perfect separation, even if they end up with terminal nodes containing only a handful of records. But will a rule founded on the behavior of two or three records work well on new data? Most likely not, so it is crucial that data miners should also take into account the support of each rule, that is, the rule's coverage or "how many cases constitute the rule."

Maximum purity and the respective high confidence scores are not the only things that data miners should consider. They should avoid models that overfit the training dataset since these models fail to provide generalizable predictions and will most probably collapse when used on unseen data. Analysts should develop stable models that capture general data patterns, and one way to avoid unpleasant surprises is to ensure that the rules' support is acceptably high. Therefore analysts should request a relatively high number of records for the model's parent and child nodes. Although the respective settings also depend on the total number of records available, it is generally recommended to keep the number of records for the terminal nodes at least above 100 and if possible between 200 and 300. This will ensure reasonably large nodes and will eliminate the risk of modeling patterns that only apply for the specific records analyzed.

Moreover, specific decision tree algorithms incorporate an integrated pruning procedure which trims the tree after full growth. The main concept behind pruning is to collapse specific tree branches in order to end up with a smaller and more stable optimal subtree with a simpler structure, an equivalent performance, but with better validation properties. Tree pruning is a useful feature offered by specific decision tree algorithms and it should always be selected where available.

Finally, as in any predictive model, the decision tree results should be validated and evaluated, at a disjoint dataset, before their actual usage and deployment.

Due to their transparency and their ability to examine the relation of many inputs to the outcome, decision trees are commonly applied in clustering to profile the revealed clusters. Moreover, they can be used as scoring models for assigning new cases to the revealed clusters using the derived rule set.

Using Decision Trees for Cluster Profiling and Updating

A prerequisite for using a decision tree model for cluster profiling and updating is that the relevant model presents a satisfactory level of accuracy in predicting all revealed clusters. Underperforming models with high overall error (misclassification) rates would complicate instead of helping the procedure.

Decision tree models can also help with the identification of records that do not fit well in the revealed clusters. These models separate records into homogeneous subsets (terminal nodes) with respect to the cluster membership field (the target field). An efficient decision tree model will successfully partition the dataset into pure subsets, dominated by a single cluster. The records that the model fails to separate land on terminal nodes with balanced proportions of multiple clusters. These records should be examined separately. The respective scoring rules would have low confidence values, indicating a large level of uncertainty in cluster assignment. Users aiming at a more solid and consistent clustering solution might decide not to use these rules for cluster scoring and classify the relevant records as an "unidentified" cluster.

Figure 3.27 shows the first two levels of a decision tree model that was built to profile the cluster membership field of our telecommunications example presented in "Understanding the Clusters through Profiling".

The number of voice calls is selected for the first split. Customers with high voice usage are further partitioned according to the total number of roaming calls.

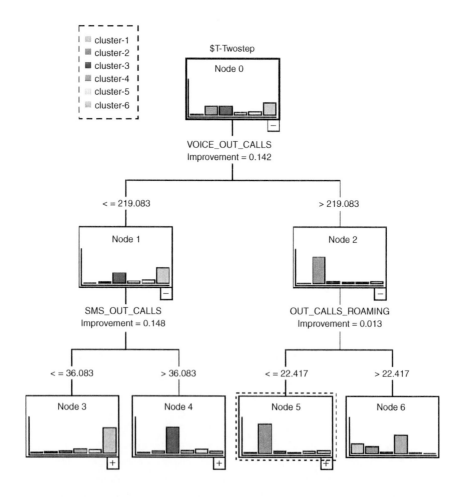

Figure 3.27 A decision tree model applied for profiling the clusters.

High roaming usage leads to the rightmost node 6 which predominantly contains members of cluster 4. On the other side of this branch, high voice and relatively lower roaming usage results in node 5 which is dominated by cluster 2. Customers with relatively lower voice usage are grouped further according to their SMS usage. Customers with high SMS usage land on node 4, in which cluster 3 is the modal category. On the left side of the branch, customers with lower SMS (and voice) usage end up in node 3 which mainly consists of members of cluster 6. Although additional partitions not shown here further refine the separation and outline the differentiating characteristics of all clusters, the fist two splits of the model have already started to portray the profile of specific clusters (clusters 4, 2, 3, and 6).

As mentioned above, behind each terminal node there is a relevant classification rule. In Figure 3.28 we present some of these rules which, as explained earlier, can score and assign new records to the revealed clusters.

Each rule is presented with the relative:

- **support** – the number of customers at the respective terminal node; and
- **confidence** – the proportion of the modal category, in this example cluster, in the respective terminal node.

By examining the rules subset presented in the screenshot of Figure 3.28, we can see that cluster 3 is mainly associated with increased SMS usage, cluster 4 with high roaming usage, and cluster 6 with relatively low usage of all services.

THE ADVANTAGES OF USING DECISION TREES FOR CLASSIFICATION MODELING

Decision trees are among the most commonly used classification techniques since they present significant advantages:

- They generate simple, straightforward, and understandable rules which provide insight into the way that predictors are associated with the output.
- They are fast and scalable and they can efficiently handle a large number of records and predictors.
- They integrate a field screening process which permits them to narrow the set of predictors by filtering out the irrelevant ones. The best predictors are selected for each partition. Predictors with marginal relevance to the output are filtered out from the model training procedure.
- They are not affected by the possible correlation between predictors and the presence of multicollinearity, a situation that can seriously affect the stability of other classification models.

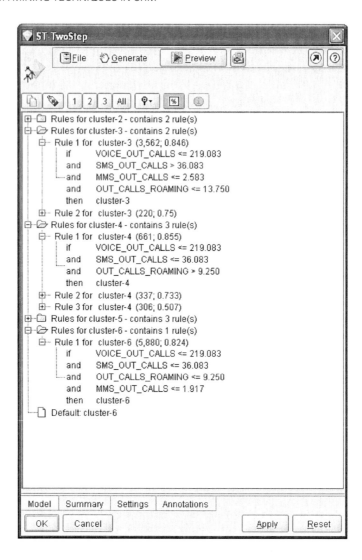

Figure 3.28 The generated decision tree rule set for cluster assignment.

- Compared to traditional statistical classification techniques (for instance, logistic regression), they have less stringent assumptions and they can provide reliable models in most of the situations.
- They can handle interactions among predictor fields in a straightforward way.
- Due to the integrated discretization (binning) of the continuous predictors, they are less sensitive to the presence of outliers and they can also capture nonlinear relationships.

On the other hand, when specific statistical assumptions are met, traditional statistical techniques can yield comparable or even better results than decision trees. Moreover, in decision trees, the model is represented by a set of rules, the number of which may be quite large. This fact may complicate understanding of the model, particularly in the case of complex and multilevel partitions. In traditional statistical techniques (like logistic regression), the inputs–output association is represented by one or a few overall equations and with respective coefficients which denote the effect of each predictor on the output.

ONE GOAL, DIFFERENT DECISION TREE ALGORITHMS: C&RT, C5.0, AND CHAID

There are various decision tree algorithms with different tree growth methods. All of them have the same goal of maximizing the total purity by identifying sub-segments dominated by a specific outcome. However, they differ according to the measure they use for selecting the optimal split.

Classification and Regression Trees (C&RT) produce splits of two child nodes, also referred to as binary splits. They typically incorporate an impurity measure named Gini for the splits. The Gini coefficient is a measure of dispersion that depends on the distribution of the outcome categories. It ranges from 0 to 1 and has a maximum value (worst case) in the case of balanced distributions of the outcome categories and a minimum value (best case) when all records of a node are concentrated in a single category.

The Gini Impurity Measure Used in C&RT

The formula for the Gini measure used in C&RT models is as follows:

$$\text{Gini} = 1 - \sum_i P(t_i)^2$$

where $P(t_i)$ is the proportion of cases in node t that are in output category i.

In the case of an output field with three categories, a node with a balanced outcome distribution of records (1/3, 1/3, 1/3) has a Gini value of 0.667. On the contrary, a pure node with all records assigned to a single category and a distribution of (1, 0, 0) gets a Gini value of 0.

The Gini impurity measure for a specific split is the weighted average of the resulting child nodes. Consequently, a split which results in two nodes of equal size with respective Gini measures of 0.4 and 0.2 has a total Gini value of 0.5°0.4 + 0.5°0.2 = 0.3. At each branch, all predictors are evaluated and the predictor that results in the maximum impurity reduction or equivalently the greatest purity improvement is selected for partitioning.

The C5.0 algorithm can produce more than two subgroups at each split, offering non-binary splits. The evaluation of possible splits is based on the ratio of the information gain, which is an information theory measure.

The Information Gain Measure Used in C5.0

Although a thorough explanation of the information gain is beyond the scope of this book, we will try to explain its main concepts. The information represents the bits needed to describe the outcome category of a particular node. The information depends on the probabilities (proportions) of the outcome classes and it can be expressed in bits which can be considered as the simple Yes/No questions that are needed to determine the outcome category. The formula for information is as follows:

$$\text{Information} = -1 * \left[\sum_i P(t_i) * \log_2(P(t_i)) \right]$$

where $P(t)_i$ is the proportion of cases in node t that are in output category i.

So, if, for example, in a node there are three equally balanced category outcomes, the node information would be $-[\log_2(1/3)]$ or 1.58 bits.

The information of a split is simply the weighted average of the information of the child nodes. Information gained by partitioning the data based on a selected predictor X is measured by:

INFORMATION GAIN (because of split on X)

= INFORMATION (PARENT NODE)

− INFORMATION (after splitting on X).

The C5.0 algorithm will choose for the split the predictor which results in the maximum information gain. In fact it uses a normalized format of the information gain (the information gain ratio) which also fixes a bias in previous versions of the algorithm toward large and bushy trees.

Both C5.0 and C&RT tend to provide bushy trees. That is why they incorporate an integrated pruning procedure for producing smaller trees of equivalent predictive performance. Analysts can specify options that control the extent and the severity of the pruning.

The CHAID algorithm is a powerful and efficient decision tree technique which also produces multiple splits and is based on the chi-square statistical test of independence of two categorical fields. CHAID stands for Chi-square Automatic Interaction Detector. In the CHAID model, the chi-square test is used to examine whether the output and the evaluated predictor are independent. At each branch, all predictors are evaluated for splitting according to this test. The most significant predictor, that is, the predictor with the smallest p-value (observed significance level) on the respective chi-square test, is selected for splitting, provided of course that the respective p-value is below a specified threshold (significance level of the test). Before evaluating predictors for splitting, the following actions take place:

1. Continuous predictors are discretized in bands of equal size, typically 10 groups of 10% each, and recoded to categorical fields with ordered categories.
2. Predictors are regrouped and categories that do not differ with respect to the outcome are merged. This regrouping of predictor categories is also based on relevant chi-square tests of independence.

In all the above models (C&RT, C5.0, CHAID), analysts can specify in advance the minimum number of records in the child nodes to ensure a minimum support level for the resulting rules.

In the context of this book we will only present the recommended options for the CHAID algorithm.

Recommended CHAID Options

In Figures 3.29–3.31 and Table 3.14 the recommended CHAID options are presented and explained for fine tuning the model in IBM SPSS Modeler and in any other data mining software which offers the specific technique.

Table 3.14   IBM SPSS Modeler recommended CHAID options.

Option	Setting	Functionality/reasoning for selection
Method	**CHAID**	This option determines the tree growth method. CHAID is the IBM SPSS Modeler default option and the tree growing method recommended to start with in most classification tasks

(*continued overleaf*)

Table 3.14   *(continued)*

Option	Setting	Functionality/reasoning for selection
		Worth trying alternatives: for example, exhaustive CHAID, a modification of the CHAID algorithm that takes longer to be trained but often gives high-quality results. Users should also try other decision tree algorithms and C5.0 in particular
Maximum tree depth/levels below root	3–6	This option determines the maximum allowable number of consecutive partitions of the data. Although this option is also related to the available number of records, users should try to achieve effective results without ending up with bushy and complicated trees For trees mainly constructed for profiling purposes, a depth of three to four levels is typically adequate For predictive purposes, a depth of five to six levels is sufficient, whereas larger trees, even when the available number of records allows them, would probably provide complicated rules that are hard to examine and evaluate
Alpha for splitting	0.05	This option determines the significance level for the chi-square statistical test used for splitting. A split is performed only if a significant predictor can be found, that is, the *p*-value of the corresponding statistical test is less than the specified alpha for splitting In plain language, this means that by increasing the alpha for the splitting value (normally up to 0.10), the test for splitting becomes less strict and splitting is made easier, resulting in potentially larger trees. A tree previously terminated due to non-significant predictors may be further grown because of loosening in the test's criteria. Lower values (normally up to 0.01) tend to give smaller trees

Table 3.14   (*continued*)

Option	Setting	Functionality/reasoning for selection
Alpha for merging	**0.05**	This option determines the significance level for the merging of predictor categories. Higher values (normally up to 0.10) hinder the merging of predictor categories and a value of 1.0 totally prevents merging. Lower values (normally up to 0.01) facilitate the collapsing of predictor categories
Minimum records in parent branch/ minimum records in child branch	**Minimum 100 records for any child branch**	These options specify the minimum allowable number of records in the parent and child nodes of the tree. The tree growth stops if the size of the parent or of the resulting child nodes is less than the specified values. The size of the parent node should always be set higher (typically two times higher) than the corresponding size of the child nodes The requested values can be expressed in terms of a percentage of overall training data or in terms of absolute number of records Although the respective settings also depend on the total number of records available, it is generally recommended to keep the number of records for the terminal (child) nodes at least above 100 and if possible between 200 and 300. Large values of these settings provide robust rules "supported" by many cases/records which we expect to perform well when used in new datasets

SUMMARY

In this chapter we focused on the modeling techniques used in the context of segmentation, PCA, and clustering techniques in particular.

PCA is an unsupervised data reduction technique usually applied in order to prepare data for clustering. It is used for effectively replacing a large set of original continuous fields with a core set of composite measures.

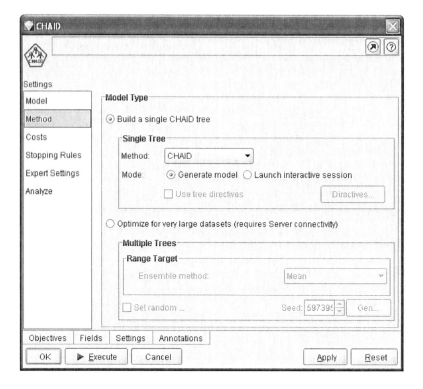

Figure 3.29 IBM SPSS Modeler recommended CHAID Model options.

Things to bear in mind about PCA:

- Numerous different criteria can be taken into account for determining the number of components to be extracted. Most of them relate to the amount of original information jointly and individually accounted by the components. The most important criterion, though, must be the interpretability of the components and whether they make sense from a business point of view.
- Component interpretation is critical since the derived components will be used in subsequent tasks. This usually includes an examination of the loadings (correlations) of the original inputs on components, preferably after applying a Varimax rotation.
- Derived component scores can be used to represent and to substitute for the original fields. Provided of course that they are derived by a properly developed model, they can be used to simplify and even refine subsequent modeling tasks, including clustering, since they equally represent all the underlying data dimensions.

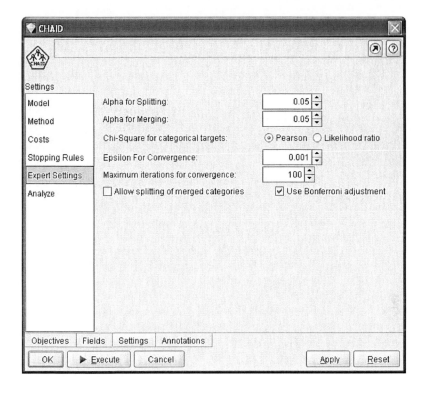

Figure 3.30   IBM SPSS Modeler recommended CHAID Expert options.

Clustering techniques are applied for unsupervised segmentation. They induce natural groupings of records/customers with similar characteristics. The revealed clusters are directed by the data and not by subjective and predefined business opinions. There is no right or wrong clustering solution. The value of each solution depends on its ability to represent transparent, meaningful, and actionable customer typologies.

Things to bear in mind about clustering:

- The clustering solution reflects the similarities and differences embedded in the specific analyzed data. Therefore, the analysis should start with the selection of the clustering fields, in other words the selection of the appropriate segmentation criteria, which will best address the defined business objective.
- The clustering solution could be biased by large differences in the measurement scales and by intercorrelations of the input fields. Therefore, a recommended approach is to firstly run PCA to identify the distinct data dimensions and then use the (standardized) component scores as inputs for clustering.

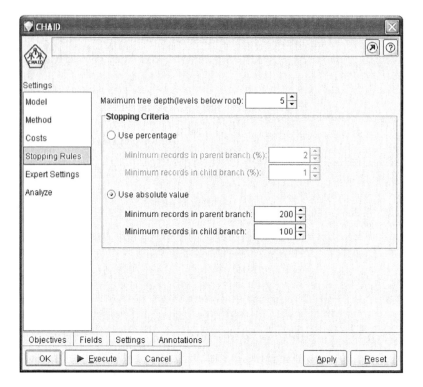

Figure 3.31 IBM SPSS Modeler recommended CHAID Stopping Criteria.

- Although specific clustering techniques offer automatic clustering methods for selection of the clusters to fit, it is recommended to experiment and try different solutions before approving the one for deployment.
- The evaluation of the clustering solution should involve an examination of the separation and of the homogeneity of the derived clusters. Derived clusters should present minimum intra-cluster and maximum inter-cluster variation. But most of all, they should be interpretable and evoke business opportunities. They should also correspond to recognizable customer "types" with clearly differentiated characteristics. A clustering solution is justified only if it makes business sense and if it can be used for the "personalized" handling of customers, for instance for developing new products or for making new offers tailored to the characteristics of each cluster. That is why data miners should work together with the involved marketers in selecting the optimal clustering solution.
- Profiling is an essential step in the evaluation of the clustering solution. By using statistics and charts, even with the use of auxiliary models, data miners should recognize the differentiating characteristics of each cluster. Each cluster

is characterized by its centroid or cluster center. This is a virtual "prototype record," the vector of the means of all clustering inputs. It is the central point of the cluster. Cluster centers can be considered as the most typical member cases of each cluster and therefore are usually examined and explored for cluster profiling.

- Apart from simple tables and charts of descriptive statistics, revealed clusters can also be profiled through the use of classification modeling techniques which examine the association of the clusters with the fields of interest. Decision trees, due to their transparency and the intuitive form of their results, are commonly applied to explore the structure of the clusters.

Table 3.15 summarizes and compares the characteristics of the clustering techniques presented in the previous sections, namely the K-means, TwoStep, and Kohonen network techniques.

Table 3.15   Comparative table of clustering techniques.

	K-means	**TwoStep**	**Kohonen network/SOM**
Methodology description	Iterative procedure based on a selected (typically Euclidean) distance measure	In the first phase records, through a single data pass, are grouped into pre-clusters. In the second phase pre-clusters are further grouped into the final clusters through hierarchical clustering	Based on neural networks. Clusters are spatially arranged in a grid map with distances indicating their similarities
Handling of categorical clustering fields	Yes, through a recoding into indicator fields	Yes	Yes, through a recoding into indicator fields
Number of clusters	Analysts specify in advance the number of clusters to fit; therefore it requires multiple runs and tests	The number of clusters is automatically determined according to specific criteria	Analysts specify the (maximum) number of output neurons. These neurons indicate the probable clusters

(continued overleaf)

Table 3.15   (*continued*)

	K-means	TwoStep	Kohonen network/SOM
Speed	Fastest	Fast	Not so fast
Integrated outlier handling (in IBM SPSS Modeler)	No	Yes	No
Recommended data preparation	Standardization and outlier handling	Standardization and outlier handling	Standardization and outlier handling

CHAPTER FOUR

The Mining Data Mart

DESIGNING THE MINING DATA MART

The success of a data mining project strongly depends on the breadth and quality of the available data. That is why the data preparation phase is typically the most time-consuming phase of the project. In the following sections we will deal with the issues of selecting, preparing, and organizing data for data mining purposes.

Data mining applications should not be considered as one-off projects but rather as continuous processes, integrated into the organization's marketing strategy. Data mining has to be "operationalized." Derived results should be made available to marketers to guide them in their everyday marketing activities. The results should also be loaded into the organization's front-line systems in order to enable "personalized" customer handling. This approach requires the setting up of well-organized data mining procedures, designed to serve specific business goals, instead of occasional attempts which just aim to cover sporadic needs.

In order to achieve this and become a "predictive enterprise" an organization should focus on the data to be mined. Since the goal is to turn data into actionable knowledge, a vital step in this "mining quest" is to build the appropriate data infrastructure. Ad hoc data extraction and queries which just provide answers to a particular business problem may soon end up as a huge mess of unstructured information. The proposed approach is to design and build a central mining data mart that will serve as the main data repository for the majority of the data mining applications. All relevant information should be taken into account in the design of the data mart. Useful information from all available data sources, including internal sources such as transactional, billing, and operational systems, and external sources such as market surveys and third-party lists, should be collected and consolidated in the data mart framework. After all, this is the main idea of the data mart: to combine all important blocks of information in a central repository that can enable the organization to have a complete a view of each customer.

The mining data mart should:

- Integrate data from all relevant sources.
- Provide a complete view of the customer by including all the attributes that characterize each customer and his or her relationship with the organization.
- Contain preprocessed information, summarized at the minimum level of interest, for instance at a product account or customer level. To facilitate data preparation for mining purposes, preliminary aggregations and calculations should be integrated into the building and updating process of the data mart.
- Be updated on a regular and frequent basis to summarize the current view of the customer.
- Cover a sufficient time period (enough days or months, depending on the specific situation) so that the relevant data can reveal stable and non-volatile behavioral patterns.
- Contain current and past data so that the view of the customer can be examined at different moments in time. This is necessary since in many data mining projects analysts have to examine historical data and analyze customers before the occurrence of a specific event, for instance before purchasing an additional product or before churning to the competition.
- Cover the requirements of the majority of the upcoming mining tasks, without the need for additional implementations and interventions from IT. The designed data mart could not possibly cover all the needs that might arise in the future. After all, there is always the possibility of extracting additional data from the original data sources or for preparing the original data in a different way. Its purpose is to provide rapid access to commonly used data and to support the most important and most common mining tasks. There is a thin line between incorporating too much or too little information. Although there is no rule of thumb suitable for all situations, it would be useful to keep in mind that raw transactional/operational data may provide a depth of information but they also slow down performance and complicate the data preparation procedure. At the other end, high-level aggregations may depreciate the predictive power hidden in detailed data. In conclusion, the data mart should be designed to be as simple as possible with the crucial mining operations in mind. Falling into the trap of designing the "mother of all data marts" will most probably lead to a complicated solution, no simpler than the raw transactional data it was supposed to replace.

In the following sections we will present mining data mart proposals for mobile telephony (consumer customers), retail banking, and retailing. Obviously, the amount of available data, the needs, and the requirements of each organization differ. Each case merits special consideration and design, which is why these proposals should only be regarded as a general framework that delineates the minimum information needed to support some of the most common data mining activities. The proposals include a set of tables and a set of informative input fields and derived indicators that have proved useful in real mining applications.

It should be noted, though, that the proposed data marts mostly focus on data from internal data sources. Data from external data sources such as market survey data are not covered in detail. Similarly, web data (data logged during a visit to the organization's web site) and unstructured text data (such as comments, requests, complaints recorded as free text during a customer's interaction with a call center agent) are also not covered in these proposals.

THE TIME FRAME COVERED BY THE MINING DATA MART

In general, a mining data mart should incorporate current and past data on all crucial customer attributes to permit their monitoring over time.

Clustering models used for behavioral segmentation are less demanding in terms of past data compared to supervised/predictive models. Although typically 6–12 months of past data are sufficient for segmentation models, the data marts proposed here cover a period of two years to also empower predictive modeling.

Behavioral segmentation models are based only on the current or, to state it more precisely, on the most recent view of the customer and require a simple snapshot of this view, as shown in Figure 4.1 below. However, since the objective is to identify a segmentation solution founded on consistent and not on random behavioral patterns, the included data should cover a sufficient time period of at least six months.

Predictive models such as cross-selling and churn models, on the other hand, require the modeling dataset to be split into different time periods. To identify data patterns associated with the occurrence of an event, the model should analyze the customer profile before the occurrence of the event. Therefore, analysts should focus on a past moment and analyze the customer view before the purchase of an add-on product or before churning to a competitor.

Let us consider, for example, a typical churn model. During the model training phase, the model dataset should be split to cover the following periods:

1. **Historical period:** Used for building the customer view in a past time period, before the occurrence of the event. It refers to the distant past and only predictors (input attributes) are used to build the customer view.
2. **Latency period:** Reserved for taking into account the time needed to collect all necessary information to score new cases, predict future churners, and execute the relevant campaigns.

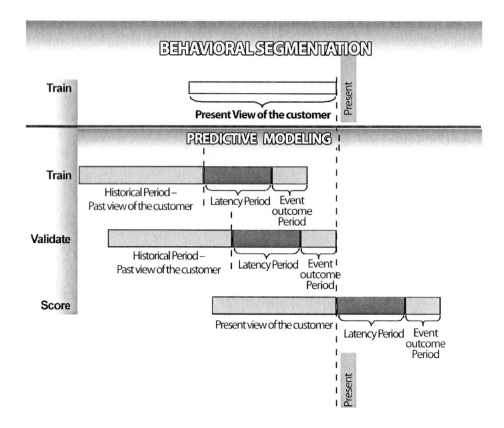

Figure 4.1   Data setup of behavioral segmentation and predictive models.

3. **Event outcome period:** Used for recording the event outcome, for example, churned within this period or not. It follows the historical and latency periods and is used to define the output field of the supervised model.

The model is trained by associating input data patterns of the historical period to specific event outcomes recorded in the event outcome period.

Typically, in the validation phase, the model's predictive performance is evaluated in a disjoint dataset which covers different time periods. In the deployment phase new cases are scored according to their present view, specifically, according to the input data patterns observed in the period just before the present. The event outcome is unknown and its future value is predicted. The whole setup is illustrated in Figure 4.1.

Therefore, a mining data mart designed to address advanced analytical needs, and predictive modeling in particular, should incorporate enough past data to model the changes of the customer view over different time periods.

THE MINING DATA MART FOR RETAIL BANKING

The main purpose of the retail banking mining data mart is to provide rapid access to pre-summarized information that can cover specific mining as well as marketing and reporting needs, including:

- Development of data mining models and procedures.
- Reporting and monitoring of KPIs.
- Extraction and evaluation of campaign lists.

A time period of two years is proposed as a time frame for the data mart in order to support most data mining applications, including segmentation and predictive modeling.

In terms of the customer attributes covered, the retail banking data mart should cover, as a minimum, the following information:

- Registration and socio-demographic information of each customer.
- Product mix (ownership) and product utilization (product balances) over the time period covered and information on the evolution of the customer's portfolio.
- Frequency (number) and volume (amount) of transactions by type (credit, debit, withdrawal, deposit, etc.) and channel (ATM, branch, web banking, etc.).
- For the specific case of credit cards, frequency and volume of purchases by type (one-off, installments, etc.) and category of merchant.
- Credit score and arrears history.
- Information on the profitability of each customer over the time period covered.
- History of applications for purchasing products, application sources, and results (approvals, declines, etc.).
- Campaign contacts and responses.
- Recorded complaints and complaint outcomes.
- Current and past status of each customer (active, inactive, dormant, etc.) and status transitions over the time period covered.
- Core segmentation information for each customer (business, affluent, mass, etc.), where available, and information on segment migrations.

The proposed data mart for retail banking which is presented below has been designed bearing in mind the minimum information requirements for the most common analytical needs. It incorporates information at a customer and product account level and covers most aspects of the customer's relationship with the bank.

The main information tables are supplemented by lookup tables (also referred to as reference tables) which group products and transactions into a manageable number of categories and facilitate further summarization of the information.

The presented data mart refers to retail customers and mainly focuses on information from internal data sources, such as core banking platforms, other transactional systems, call center platforms, and other operational systems. Although most of the contained data are supposed to be extracted from the data warehouse, the data mart should be designed and constructed as a supplement to and not as a replacement for the data warehouse. Therefore, all data stored in the data mart should also reside in the data warehouse to ensure the consistency and integrity of the information.

CURRENT INFORMATION

Current data mart tables contain information on current and past customers and only include the most recent update of the relevant information. The records in this type of table are updated to reflect the latest information and are not deleted when a relationship terminates. For example, a current table with demographic information contains the latest update on such data for all customers, even those who have terminated their relationship with the bank (ex-customers). This type of table might also be used to track the status of products (openings/closings).

Customer Information

The first proposed table ("C_Customer"), Table 4.1, builds in the basic customer information, including demographic and contact details data. This table is at a customer level. Thus, each row is uniquely identified by the corresponding customer ID field which constitutes the primary key (PK) of the table. All customers that ever had a relationship with the bank are included and ex-customers are not deleted. The table contains the most recent updates of the customer information as well as the date of each record update.

Product Status

Table 4.2 records detailed information about the status of the products (product accounts) that each customer currently has or used to have in the past. Therefore it should contain all critical dates including opening, expiration, and closing dates. The first product opening can be used to calculate the tenure of the customer. If all product accounts of a customer are closed, a full churned customer is indicated. The contained information can be used to calculate the tenure of each product and

Table 4.1   Basic customer information table.

Field name	Field type	Field description
Household_ID	**Int**	**Household identification number**
Customer_ID	Int	Customer identification number (primary key)*
VAT_Num	**Char**	**VAT number**
ID_Num	Char	ID number
ID_Type_Code	**Int**	**ID type – code number**
Gender_Code	Int	Gender – code number
Marital_Status_Code	**Int**	**Marital status – code number**
Children_Num	Int	Number of children
House_Owner	**Bit**	**House owner flag**
Educational_Status_Code	Int	Educational status – code number
Occupation_Code	**Int**	**Occupation – code number**
Annual_Income_Code	Int	Annual income category – code number
Birth_Date	**Datetime**	**Date of birth**
Death_Date	Datetime	Date of death, otherwise a control date
Registration_Date	**Datetime**	**First registration date**
Registration_Branch	Int	First registration branch
Closure_Date	**Datetime**	**Date of customer relationship closure**
Closure_Branch	Int	Branch of customer relationship closure
First_Name	**Char**	**First name**
Last_Name	Char	Last name
Middle_Name	**Char**	**Middle name**
Home_Street	Char	Home street name
Home_Num	**Char**	**Home street number**
Home_City	Char	Home city
Home_State	**Char**	**Home state**
Home_ZIP	Char	Home ZIP code
Home_Country	**Char**	**Home country**
Work_Company	Char	Name of company where customer works
Work_Street	**Char**	**Work street name**
Work_Num	Char	Work street number
Work_City	**Char**	**Work city**
Work_State	Char	Work state
Work_ZIP	**Char**	**Work ZIP code**
Work_Country	Char	Work country
Home_Phone	**Char**	**Home phone number**
Mobile_Phone	Char	Mobile phone number
Work_Phone	**Char**	**Work phone number**
Home_Email	Char	Home e-mail
Work_Email	**Char**	**Work e-mail**
Last_Update	Datetime	Last date of customer record update

Customer_ID is the primary key (PK) of this table.

Table 4.2 The product status table.

Field name	Field type	Field description
Customer_ID	**Int**	**The customer identification number**
Account_ID	Int	The product account identification number
Primary_Flag	**Bit**	**A flag indicating if the customer ID is the primary customer for the specific account**
Product_Code	Int	The product code (see next "L_Product_Code" table)
Application_ID	**Int**	**The application identification number related to the account**
Opening_Date	Datetime	The opening date of the product account
Closing_Date	**Datetime**	**The closing date of the product account (a future control date is used when the account is still open)**
Expiration_Date	Datetime	The expiration date of the product account (a future control date is used when not applicable)
Closing_Reason	**Int**	**The recorded closing reason for closed accounts**

Customer_ID and Account_ID are the primary keys (PKs) of this table.

customer, each account's current status (open, closed), and the time of closing in the case of terminated accounts (partial churn). Similarly, a list of recent account activations and closings (e.g., within the previous month) and a list of recently acquired or churned customers can be easily derived from this table of product applications.

Another current table of important customer information is the table. This table should keep a record of all applications of customers for buying new products along with the final outcome of each application (approved, canceled, declined, etc.).

MONTHLY INFORMATION

These tables summarize data on a monthly basis. They include aggregated data that record main customer attributes over time.

This kind of table is recommended for those characteristics that should be monitored over time, such as product ownership and utilization (product balances), frequency and volume of transactions, segment and group membership of the customers. A monthly level of aggregation is proposed as a good balance between simplicity and efficiency/flexibility. Nevertheless, according to each organization's specific needs, available resources, and required detail of information, an aggregation at a lower time period (weekly, daily, etc.) could be implemented for all or for a subset of key attributes. For example, a daily tracking of credit and debit transactions could be useful for both reporting and triggering campaigns.

Segment and Group Membership

A bank typically categorizes its customers according to their characteristics into broad, distinct groups such as retail, corporate, private banking, and so on. Customers are also assigned to core segments (affluent, mass, etc.), grouped according to their status (active, dormant, etc.), and flagged according to other important attributes such as having a salary account at the bank or belonging to the bank's staff. These customer categorizations enclose important information for marketing and analytical purposes.

The first monthly table proposed ("M_Segments"), Table 4.3, incorporates information about the bank's main segmentation schemes and groupings. The contained data are at a customer level and indicate the group memberships of each customer, at the end of each month.

Product Ownership and Utilization

Any bank should monitor the product mix and product utilization of its customers. This information is tracked by the next proposed table ("M_Products"), Table 4.4, which includes monthly aggregated data on product ownership and balances. It is a table at the customer and product account level that only refers to open accounts, even if these accounts have zero balance. It summarizes product balances for each month, including both outstanding (end of month) and (daily) average balances. By mining the historical data, analysts can reveal any shifts or declines in the customer's relationship with the bank and understand the evolution of each customer's product portfolio.

A lookup table, to be presented below, can be used to group product codes and to summarize customers with respect to a higher hierarchy level of product categorization (deposits, investments, loans, etc.). This table also includes other product-specific information such as the credit limit and the minimum payment amount for credit card accounts or the approved amount in the case of loans.

Bank Transactions

Customers differ in terms of the volume and frequency of the transactions they make. They also prefer different channels for their transactions and make different

Table 4.3 The segment and group membership table.

Field name	Field type	Field description
Year_Month	**Int**	**The number that indicates the year and month of data**
Customer_ID	Int	The customer identification number
Banking_Code	**Int**	**The basic banking code that categorizes customers as retail, corporate, private banking, and so on**
VIP_Flag	Bit	A flag that indicates if a customer is considered as a VIP by the bank
Contact_Flag	**Bit**	**A flag that indicates if a customer agrees to be contacted by the bank**
Person_Flag	Bit	A flag that indicates if a customer is a person or a legal entity
Dormant_Flag	**Bit**	**A flag that indicates, in accordance with the bank's relevant definition, a dormant customer, for instance a customer with no transactions for the last 12 months**
Active_Flag	Bit	A flag that separates active from inactive customers, according to the bank's definition. For instance, a customer with null or very low total balances could be considered as inactive
Payroll_Flag	**Bit**	**A flag that indicates if a customer has a payroll account at the bank**
Segment_Code	Int	A code that allocates customers to specific segments such as affluent (customers with the highest assets, namely savings and investment balances), mass (customers with lower assets), staff (belonging to the organization's staff), and so on

Year_Month and Customer_ID are the primary keys (PKs) of this table.

types of transactions (deposits, withdrawals, etc.). The proposed table of transactions ("M_Transactions"), Table 4.5, captures all information related to customer transactional behavior. It contains monthly aggregations which summarize the number and amount of transactions per transaction type and channel, at a customer and product account level. The data recorded enable the identification of transactional patterns and the transactional profiling of the customers. Two lookup tables, to be presented below, group transaction channels and transaction codes into broader categories and facilitate further aggregation and summarization of the tables' records.

Table 4.4   The product ownership and utilization table.

Field name	Field type	Field description
Year_Month	**Int**	**The number that indicates the year and month of data**
Customer_ID	Int	The customer identification number
Account_ID	**Int**	**The product account identification number**
Primary_Flag	Bit	Primary owner of the account – flag
Product_Code	**Int**	**The account product code (see next "L_Product_Code" lookup table)**
Interest_Rate	Real	The rate of interest (when applicable)
Interest_Type	**Int**	**The type of interest (floating, fixed – when applicable)**
Credit_Limit	Real	The credit limit (when applicable – credit cards)
Fee	**Real**	**The fee amount (when applicable – credit cards)**
Approved_Amount	Real	The approved amount (when applicable – personal loans)
Used_Amount	**Real**	**The actually used amount (when applicable – personal loans disbursal)**
Installment_Amount	Real	The monthly installment amount (when applicable)
Min_Payment_Amount	**Real**	**The minimum payment amount (when applicable – credit cards)**
Outstanding_Balance	Real	The outstanding balance (end of month balance)
Average_Balance	**Real**	**The (daily) average balance**
Max_Balance	Real	The maximum balance within a month
Min_Balance	**Real**	**The minimum balance within a month**
Revenue	Real	The projected monthly revenue
Cost	**Real**	**The projected monthly cost**
Arrears_Months	Int	Months in arrears (when applicable – credit cards, personal loans, etc.)

Year_Month, Customer_ID, and Account_ID are the primary keys (PKs) of this table.

LOOKUP INFORMATION

A lookup table contains entries (records) which store key–value pairs. Each key is mapped to a corresponding value(s). The value provides additional information, such as key description or categorization information.

Table 4.5 The transactions table.

Field name	Field type	Field description
Year_Month	**Int**	**The number that indicates the year and month of data**
Customer_ID	Int	The customer identification number
Account_ID	**Int**	**The product account identification number**
Transaction_Code	Int	The transaction code (see next "L_Transaction_Code" lookup table)
Transaction_Channel	**Int**	**The transaction channel (see next "L_Transaction_Channel" lookup table)**
Num_Transactions	Int	Total number of transactions within month
Amount_Transactions	**Real**	**Total amount of transactions within month**
Min_Amount	Real	Minimum amount of transactions within month
Max_Amount	**Real**	**Maximum amount of transactions within month**

Year_Month, Customer_ID, Account_ID, Transaction_Code, and Transaction_Channel are the primary keys (PKs) of this table.

A dictionary is a typical example of a lookup table, where the key is a word and the value is the dictionary definition of that word. Similarly, if the lookup table records the marital status of a customer, the keys might be the marital status codes, for instance (1, 2, 3), and the values might be the corresponding descriptions, for instance (single, married, divorced).

Obviously, in a well-designed data mart many different lookup tables are required. An indicative list of lookup tables useful for the retail banking data mart is presented in the next sections.

Product Codes

Although the structure of the proposed data mart tables permits the extraction of information at a product code level, most data mining tasks would typically require less detailed information, such as data summaries at a product group/subgroup level.

A lookup table (such as the "L_Product_Code" table presented in Table 4.6) can be used to define and store a multilevel grouping hierarchy of product codes, mapping each product to the corresponding product groups and subgroups. When

Table 4.6   A lookup table for product codes.

Field name	Field type	Field description
Product_Code	**Int**	**The product code**
Product_Desc	Char	The product description
Financial_Group	**Char**	**The financial group of products (assets/ liabilities)**
Banking_Group	Char	The banking group of products (funds, bank assurance, loans)
Product_Group	**Char**	**The basic product group (see the categorization of table 4.7, first level)**
Product_Subgroup	Char	The product subgroup (see the categorization of table 4.7, second level)

Product_Code is the primary key (PK) of this table.

joined with other data mart tables, the lookup table enables aggregation of the respective data at the selected level of product grouping.

A typical list of product groups and subgroups that can be used for the categorization of product codes is given in Table 4.7.

Transaction Channels

The next proposed table ("L_Transaction_Channel"), Table 4.8, refers to the different transaction channels, providing a detailed description of each channel code.

The transaction channels would typically include those listed in Table 4.9.

Transaction Types

The last lookup table ("L_Transaction_Code"), Table 4.10, covers the transaction codes and introduces a corresponding multilevel grouping scheme according to the transaction type. It can be used to consolidate transactions of the same type.

Transactions are grouped into two main groups according to their kind, namely debit and credit transactions. In Table 4.10, a relevant flag (binary or dichotomous) field separates credit from debit transactions and can be used for aggregating the data accordingly.

An illustrative categorization of transactions into transaction types, groups, and subgroups is given in Table 4.11.

Table 4.7   A typical list of product groups and subgroups in retail banking.

Product group	Product subgroup
Deposits/savings	• Saving accounts • Current accounts • Notice accounts • Collaterals • Other accounts
Time deposits	• Time deposits
Investments	• Stock funds • Bond funds • Money market funds • Balanced funds • Bonds • Stocks • Other investments
Insurances	• Life assurance • Pension insurance • Health insurance • General insurance
Business products	• Business loans • Business accounts • Business services
Mortgages and home equity loans	• Mortgage • Home equity loans • Other home loans
Consumer loans	• Personal lines • Personal loans • Auto loans • Student loans • Debt consolidation • Other consumer loans
Cards	• Charge cards • Debit cards • Credit cards • Prepaid cards • Other cards

Table 4.8   A lookup table for transaction channels.

Field name	Field type	Field description
Transaction_Channel	**Int**	**The transaction channel code**
Transaction_Channel_Desc	Char	The transaction channel description

Transaction_Channel is the primary key (PK) of this table.

Table 4.9 A typical list of transaction channels in retail banking.

Transaction channels
Branch
ATM
APS (Automatic Payment System)
Internet
Phone
SMS
Standing order
Other

Table 4.10 A lookup table for transaction codes.

Field name	Field type	Field description
Transaction_Code	**Int**	**The transaction code**
Transaction_Desc	Char	The transaction description
Debit_Credit	**Bit**	**Debit/credit transaction flag**
Transaction_Type	Char	The basic transaction type (see the categorization of table 4.11, first level)
Transaction_Group	**Char**	**The transaction group (see the categorization of table 4.11, second level)**
Transaction_Subgroup	Char	The transaction subgroup (see the categorization of table 4.11, third level)

Transaction_Code is the primary key (PK) of this table.

Grouping Transactions with Other Criteria

Obviously, the transactional data can also be aggregated according to any other grouping criteria to address other mining objectives. For instance, the summarization of credit card purchase transactions by merchant type, provided of course the merchant code and type are recorded, can reveal the general purchasing habits of customers and facilitate the profiling of credit card holders by their purchase preferences.

Table 4.11   An indicative grouping hierarchy of transactions according to their type in retail banking.

Transaction type	Transaction group	Transaction subgroup
Deposits	• Deposit internal • Deposit external	• Deposit internal • Deposit external
Credit cards	• Cash advance • Purchases	• Cash advance • One-off purchases • New installments (installment due to a new purchase) • Old installments (installment due to an old purchase)
Withdrawals	• Withdrawal classic • Withdrawal overdraft	• Withdrawal classic • Withdrawal overdraft
Payments	• Banking payment • Bill payment	• Home loan payment • Business loan payment • Consumer loan payment • Credit card payment • Insurance payment • Public service payment • Telecommunication services payment • Internet services payment • Pay TV services payment • Other payment
Transfers	• Transfer internal • Transfer external	• Transfer internal • Transfer external
Queries	• Queries	• Queries

THE CUSTOMER "SIGNATURE" – FROM THE MINING DATA MART TO THE MARKETING CUSTOMER INFORMATION FILE

Most data mining applications require a one-dimensional flat table, typically at the customer level. Therefore, analysts should move one step beyond, since even the data mart tables are not immediately ready for mining. A recommended approach is to consolidate the mining data mart information into one table which should be designed to cover the key mining as well as marketing and reporting needs. This table, also referred to as the marketing customer information file or MCIF, should integrate all the important customer information, including demographics, usage, and revenue data, thus providing a customer "signature": a unified view of the customer. Through extensive data processing, the data retrieved from the data mart should be enriched with derived attributes and informative KPIs to summarize all aspects of the customer's relationship with the organization.

The MCIF typically includes data at a customer level, though the structure of the data mart presented above also supports the building of a relevant file at a different level, such as at a product account level for instance. It should be updated on a regular basis to provide the most recent view of each customer and it should be stored to track the customer view over time. The main idea is to compose a good starting point for the rapid creation of an input file for all key future analytical applications.

Creating the MCIF through Data Processing

The MCIF can be built and maintained by using:

- **Database views** created in the existing relational database management system, such as ORACLE, SQL Server, etc.
- **Procedures** created using data mining tools such as IBM SPSS Modeler (formerly Clementine) streams.

Nevertheless, the consolidation of all the available information in a single table is a challenging and demanding job which usually requires complicated data processing in order to select, transform, and integrate all relevant data. Some of the most useful data management operations and the corresponding database functions typically applied for the construction of the MCIF are as follows:

- **Select:** Used to select or discard a subset of records from a data file based on a logical condition. In SQL this is expressed as: select <fields> from <data file> where <condition>
- **Merge/join:** Matches records and adds fields from multiple data sources using key fields (for instance, the customer ID field). Several types of joins are available, including inner join, full outer join, partial outer join, and anti-join.
- **Aggregate/group by:** Aggregates multiple records based on a key field and replaces the input records with summarized output records. Several summary statistics can be applied, such as: sum, mean, min, max, standard deviation. In SQL this is expressed as: select <key fields>, <summarized fields> from <data file> group by <key fields>
- **Insert/append:** Insert is used to concatenate sets of records. It is useful when the aim is to combine data sources with the same structure (fields) but with different records. In SQL this is expressed as: insert into <data file>, <fields> select <fields> from <source data file>
- **Update:** Update is used for filling specific fields with calculated values. It can be applied to all values of a field or to the values that satisfy a logical condition.

In SQL this is expressed as: update <table> set <field> = <sub query or formula>

- **Restructure:** Pivots data and generates multiple fields based on the values of another field. Restructure is usually followed by aggregation. For instance, restructure can be applied when we have groups of related records (e.g., multiple records denoting the transaction amount and the transaction channel for each customer) that we want to rearrange so that the original multiple records are pivoted into a single record with multiple separate fields: that is, one record for each customer with multiple fields, denoting the relative sum for each distinct transaction channel. In SQL this is expressed as: case when <formula> then <value1> else <value2> end

- **Derive/compute:** Modifies data values and constructs new fields from existing ones, based on calculations, logical conditions, and functions. During lengthy data mining projects, it is common to perform several derivations, such as flags, ratios, percentages, etc.

As a simplified example of a typical data processing procedure required for the construction of the MCIF, let us consider the transaction data in Table 4.12 for two bank customers. The data denote the volume of transactions (Amount_ Transactions) per channel, in a format similar to the data mart's "M_Transactions" table presented earlier. The objective is to turn these data into informative indicators of the channel utilization for each customer.

This information is detailed and informative, yet not immediately suitable for mining purposes. It has to go through extensive data preparation before revealing the transactional profile of each customer. The data preparation steps necessary in order to summarize this information at a customer level for the needs of the MCIF include:

1. **Restructure and aggregation of the transactional data at a customer level:** The multiple records per customer (one record per year/month, account ID, transaction code, and transaction channel) are pivoted and aggregated so that each customer is represented by a single record and a set of relevant fields which indicate the total volume for each transaction channel. We assume here that no differentiation in terms of product code or time period is necessary for the needs of our analysis. The intermediate table which is produced through restructure and aggregation is given in Table 4.13.

   The IBM SPSS Modeler Restructure node used for pivoting the data is displayed in Figure 4.2.

2. **Derive new fields to enrich the original information:** After the data have been pivoted and aggregated at a customer level, the next step is to enrich the

Table 4.12 A subset of the "M_Transactions" table recording the channel utilization for two bank customers.

Year_Month	Customer_ID	Account_ID	Transaction_ Code	Transaction_ Channel	Transaction_ Channel_Desc	Amount_ Transactions
200901	1001	10011	101	1	BRANCH	100
200901	1001	10011	101	2	ATM	30
200901	1001	10012	101	2	ATM	80
200901	1001	10012	104	1	BRANCH	70
200901	1001	10012	102	1	BRANCH	90
200901	1002	10021	101	3	INTERNET	120
200901	1002	10021	101	1	BRANCH	150
200901	1002	10022	103	3	INTERNET	140
200901	1002	10022	105	3	INTERNET	180
200901	1002	10022	105	4	PHONE	120

Table 4.13   The transactional dataset after restructuring and aggregating data at the customer level.

Customer_ ID	Total transactions amount	ATM transactions amount	BRANCH transactions amount	INTERNET transactions amount	PHONE transactions amount
1001	370	110	260	0	0
1002	710	0	150	440	120

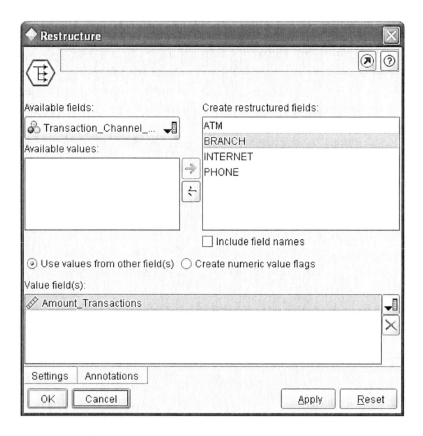

Figure 4.2   The IBM SPSS Modeler Restructure node for pivoting transactional data.

original information with a series of key indicators. For instance, by dividing the volume of each channel by the total volume of all transactions we can produce the relative volume ratio (proportion) attributable to each channel. The IBM SPSS Modeler Derive node used for this operation is displayed in Figure 4.3. The divide operation is applied to each channel volume field by using the "@" operator. The conditional ("If") part of the derive node deals with the

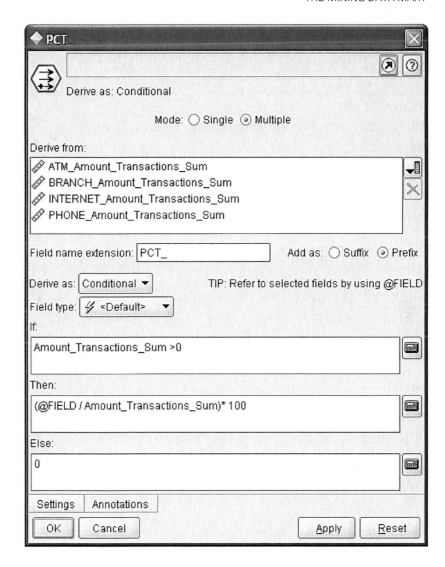

Figure 4.3   The IBM SPSS Modeler Multiple Derive node for calculating the relative utilization of each channel.

customers with zero transactions in the examined period. These customers are given a zero ratio for all channels.

In a similar manner, customers are further summarized with additional derived fields, including:

(a) A flag (computed by a conditional derive) which indicates whether a customer has used an alternative channel (Internet, phone, etc.). This field is considered as an indicator of the penetration of new channels.

Table 4.14 The derived fields concerning channel utilization.

Customer_ ID	...	% ATM transactions	% BRANCH transactions	% INTERNET transactions	% PHONE transactions	Number of different channels	Flag of alternative channels
1001	...	30	70	0	0	2	F
1002	...	0	21	62	17	3	T

(b) A numeric field measuring the number of different channels used by each customer. This field is considered as a measure of the diversity of channel utilization.

The additional information derived is presented in Table 4.14.

3. **Merge the channel utilization information with the rest of the prepared data to form the MCIF.** The information for the channel utilization joins the rest of the prepared data (customer demographics, product balances, type of transactions made, etc.) to form the MCIF. The MCIF outlines the "signature" for each customer.

Derived Measures Used to Provide an "Enriched" Customer View

The MCIF should be designed to cover the key analytical and marketing requirements of the organization; therefore its contents will vary according to the specific objectives set by the marketers. Apart from customer demographics, a MCIF incorporates simple and composite derived measures which summarize customer behavior. Typically, those derived measures include:

- **Sums/averages:** Behavioral fields are summed or averaged to summarize usage patterns over a specific time period. For instance, a set of fields denoting the monthly average number of transactions by transaction channel (ATM, Internet, phone, etc.) or by transaction type (deposit, withdrawal, etc.) can reveal the transactional profile of each customer. Sums and particularly (monthly) averages take into account the whole time period examined and not just the most recent past, ensuring the capture of stable, non-volatile, behavioral patterns.
- **Ratios:** Ratios (proportions) can be used to denote the relative preference/importance of usage behaviors. As an example let us consider the relative volume of transactions (percentage) per transaction channel. These percentages reveal the channel preferences while also adjusting for the total volume of transactions of each customer. Other examples of this type of measures include the relative usage of each call service type (voice, SMS, MMS, etc.) in

telecommunications or the relative ratio of assets vs. lending balances in banking. Apart from ratios, plain numeric differences can also be used to compare usage patterns.

- **Flags:** Flag fields are binary (dichotomous) fields that directly and explicitly record the occurrence of an event of interest or the existence of an important usage pattern. For instance, a flag field might show whether a bank customer has made any transactions through an alternative channel (phone, Internet, SMS, etc.) in the period examined or whether a telecommunications customer has used a service of particular interest.
- **Deltas:** Deltas capture changes in the customer's behavior over time. They are especially useful for supervised (predictive) modeling since a change in behavior may signify the occurrence of an upcoming event such as churn, default, and so on. Deltas can include changes per month, per quarter, or per year of behavior. As an example let us consider the monthly average amount of transactions. The ratio of this average over the last three months to the total average over the whole period examined is a delta measure which shows whether the transactional volume has changed during the most recent period.

THE MCIF FOR RETAIL BANKING

Although the MCIF's contents reflect the priorities and the specific mining goals of each organization, in Table 4.15 we try to present an indicative listing of attributes, mostly related to product ownership and transactional behavior, which illustrates the main data dimensions typically used for mining needs in retail banking.

Table 4.15 An indicative listing of the contents of the MCIF for retail banking.

Field_Name	Field description
Customer socio-demographics:	
Customer_ID	Customer identification number
Gender_Code	Gender
Marital_Status	Marital status
Children_Num	Number of children
House_Owner	House owner flag
Educational_Status	Educational status
Occupation	Occupation category
Annual_Income	Annual income band
Birth_Date	Date of birth

(continued overleaf)

Table 4.15   (*continued*)

Field_Name	Field description
Registration_Date	First registration date
Home_ZIP	Home postal code
Customer segment and group information:	
Banking_Code	The basic banking code that categorizes customers as retail, corporate, private banking, etc., customers
VIP_Flag	A flag that indicates if a customer is considered as a VIP by the bank
Person_Flag	A flag that indicates if a customer is a person or a legal entity
Dormant_Flag	A flag that indicates, in accordance with the bank's relevant definition, a dormant customer, for instance a customer with no transactions for the last 12 months
Active_Flag	A flag that separates active from inactive customers, according to the bank's definition. For instance, a customer with null or very low total balances could be considered as inactive
Payroll_Flag	A flag that indicates if a customer has a payroll account in the bank
Segment_Code	A code that allocates customers to specific segments such as affluent (customers with the highest assets, namely savings and investment balances), mass (customers with lower assets), staff (belonging to the organization's staff), and so on.
Product ownership:	
Saving_Accounts_Count	Number of saving accounts
Current_Accounts_Count	Number of current accounts
Time_Deposits_Count	Number of time deposit accounts
Stock_Funds_Count	Number of stock funds
Bond_Funds_Count	Number of bond funds
Money_Market_Funds_Count	Number of money market funds
Balanced_Funds_Count	Number of balanced funds
Bonds_Count	Number of bonds
Stocks_Count	Number of stocks
Life_Assurance_Count	Number of life insurance contracts
Pension_Insurance_Count	Number of pension insurance contracts
Health_Insurance_Count	Number of health insurance contracts
General_Insurances_Count	Number of general insurance contracts
Business_Loans_Count	Number of business loans
Business_Accounts_Count	Number of business accounts
Business_Services_Count	Number of business services
Mortgage_Count	Number of mortgage loans

Table 4.15   *(continued)*

Field_Name	Field description
Home_Equity_Loans_Count	Number of home equity loans
Personal_Lines_Count	Number of personal lines
Personal_Loans_Count	Number of personal loans
Auto_Loans_Count	Number of auto loans
Student_Loans_Count	Number of student loans
Debt_Consolidation_Count	Number of debt consolidation loans
Charge_Cards_Count	Number of charge cards
Debit_Cards_Count	Number of debit cards
Credit_Cards_Count	Number of credit cards
Prepaid_Cards_Count	Number of prepaid cards
Product balances:	
Saving_Accounts_Balance	Total balance of saving accounts
Current_Accounts_Balance	Total balance of current accounts
Time_Deposits_Balance	Total balance of time deposit accounts
Stock_Funds_Balance	Total balance of stock funds
Bond_Funds_Balance	Total balance of bond funds
Money_Market_Funds_Balance	Total balance of money market funds
Balanced_Funds_Balance	Total balance of balanced funds
Bonds_Balance	Total balance of bonds
Stocks_Balance	Total balance of stocks
Life_Assurance_Balance	Total balance of life insurance contracts
Pension_Insurance_Balance	Total balance of pension insurance contracts
Health_Insurance_Balance	Total balance of health insurance contracts
General_Insurances_Balance	Total balance of general insurance contracts
Business_Loans_Balance	Total balance of business loans
Business_Accounts_Balance	Total balance of business accounts
Business_Services_Balance	Total balance of business services
Mortgage_Balance	Total balance of mortgage loans
Home_Equity_Loans_Balance	Total balance of home equity loans
Personal_Lines_Balance	Total balance of personal lines
Personal_Loans_Balance	Total balance of personal loans
Auto_Loans_Balance	Total balance of auto loans
Student_Loans_Balance	Total balance of student loans
Debt_Consolidation_Balance	Total balance of debt consolidation loans

(continued overleaf)

Table 4.15 (*continued*)

Field_Name	Field description
Charge_Cards_Balance	Total balance of charge cards
Debit_Cards_Balance	Total balance of debit cards
Credit_Cards_Balance	Total balance of credit cards
Prepaid_Cards_Balance	Total balance of prepaid cards
Transaction types – frequency and volume:	
Deposit_Internal_Num	Number of deposit transactions to own accounts
Deposit_External_Num	Number of deposit transactions to other accounts
Withdrawal_Num	Number of withdrawal transactions
Cash_Advance_Num	Number of cash advance transactions
Payment_Credit_Cards_Num	Number of credit cards payment transactions
Payment_Consumer_Loans_Num	Number of consumer loans payment transactions
Payment_Mortgage_Loans_Num	Number of mortgage loans payment transactions
Payment_Business_Loans_Num	Number of business loans payment transactions
Payment_Insurances_Num	Number of insurances payment transactions
Payment_Public_Services_Num	Number of public services payment transactions
Payment_Telecommunications_Num	Number of telecommunication providers payment transactions
Payment_Internet_Num	Number of Internet providers payment transactions
Payment_Pay_TV_Num	Number of pay TV payment transactions
Payment_Other_Num	Number of other payment transactions
Transfer_Internal_Num	Number of transfer transactions to own accounts
Transfer_External_Num	Number of transfer transactions to other accounts
Deposit_Internal_Amount	Amount of deposit transactions to own accounts
Deposit_External_Amount	Amount of deposit transactions to other accounts
Withdrawal_Amount	Amount of withdrawal transactions
Cash_Advance_Amount	Amount of cash advance transactions
Payment_Credit_Cards_Amount	Amount of credit cards payment transactions
Payment_Consumer_Loans_Amount	Amount of consumer loans payment transactions
Payment_Mortgage_Loans_Amount	Amount of mortgage loans payment transactions
Payment_Business_Loans_Amount	Amount of business loans payment transactions
Payment_Insurances_Amount	Amount of insurances payment transactions
Payment_Public_Services_Amount	Amount of public services payment transactions
Payment_Telecommunications_Amount	Amount of telecommunication providers payment transactions
Payment_Internet_Amount	Amount of Internet providers payment transactions
Payment_Pay_TV_Amount	Amount of pay TV payment transactions
Payment_Other_Amount	Amount of other payment transactions

Table 4.15   (*continued*)

Field_Name	Field description
Transfer_Internal_Amount	Amount of transfer transactions to own accounts
Transfer_External_Amount	Amount of transfer transactions to other accounts
Transaction channels – frequency and volume:	
Branch_Num	Number of branch transactions
ATM_Num	Number of ATM transactions
APS_Num	Number of APS transactions
Internet_Num	Number of Internet transactions
Phone_Num	Number of phone transactions
SMS_Num	Number of SMS transactions
Standing_Order__Num	Number of standing orders transactions
Branch_Amount	Amount of branch transactions
ATM_Amount	Amount of ATM transactions
APS_Amount	Amount of APS transactions
Internet_Amount	Amount of Internet transactions
Phone_Amount	Amount of phone transactions
SMS_Amount	Amount of SMS transactions
Standing_Order_Amount	Amount of standing orders transactions
Financial and bucket information:	
Total_Revenue	Total revenue projected by customer
Total_Cost	Total cost projected by customer
Max_Bucket_Status	Maximum bucket status from loans

Examples of Derived Measures Used to Enrich the MCIF

The MCIF can also be enriched with averages, flags, ratios, deltas, and other more sophisticated derived fields which summarize the customer's behavior. An indicative list of more complex KPI examples is as follows:

- Ratio of total savings to total loans (balances).
- Percentage (%) of high-risk investments: high-risk investments as a percentage of total investments (balances).
- Ratio of credit card payments to credit card balances.
- Percentage (%) of credit limit or approved amount actually used (for credit cards and loans respectively).

- Number of months with minimum savings balances below a specific threshold, for instance <$50.
- Number of months with bucket >1 during the examined period.
- Maximum bucket during the examined period.
- Savings balances delta. Compares the savings balances of the most recent period, for example, the last three months, to those of the previous three months and denotes the respective percentage difference.
- Flag of usage of alternative channels (phone, Internet, SMS, etc.).
- Percentage (%) of teller (branch) transactions: teller transactions' amount as a percentage of total transactions.
- Number of different product categories that each customer owns.
- Number of different channels used by each customer.

THE MINING DATA MART FOR MOBILE TELEPHONY CONSUMER (RESIDENTIAL) CUSTOMERS

The main principles for the design of an efficient mining data mart presented in the previous sections also apply in the case of mobile telephony. Therefore, the mining data mart should contain consolidated and pre-summarized information tailored to cover the majority of the analytical needs. Its purpose is to provide rapid access to pre-aggregated data and to support the key analytical and marketing tasks without the need for additional implementations and interventions from IT.

The selected level of aggregation should ensure reduced query execution times and good performance without sacrificing important information.

The data mart should integrate data from all relevant data sources in order to cover all the information elements required for analyzing and predicting customer behavior. Therefore it should incorporate:

- Traffic information, such as number and minutes/volume of calls by call type, source/destination network, and so on. This information is stored in CDRs (Call Detail Records) providing a complete usage profile of each customer.
- Information about the customers and the line numbers (i.e., MSISDNs).

More specifically, it should cover, at a minimum, the following data dimensions:

- Registration and socio-demographic information of each customer.
- Core segmentation information (consumer postpaid/contractual, consumer prepaid customers) and information on segment migrations.

- Outgoing traffic to identify outgoing usage patterns (number and traffic by call and destination type) over time.
- Incoming traffic to identify incoming usage patterns (number and traffic by call and origin type) over time.
- Financial information such as profit and cost for each customer (billed revenue, interconnection cost, incoming revenue, etc.).
- Purchases of handsets and accessories (e.g., hands free, Bluetooth, etc.).
- Value added services (VAS) such as news, music, etc.
- Campaign contacts and responses.
- Recorded complaint outcomes and results.
- Billing, payment, and credit history (average number of days till payment, number of times in suspension, average time remaining in suspension, etc.).

A period of two years is proposed as the time frame for the data mart in order to support advanced data mining applications.

Finally, the data mart should be updated on a continuous basis to include the most recent view of each customer. The database diagram in Figure 4.4 shows an example of an indicative schema which depicts the established relationships between the tables of CDRs, customers and contracts, outlining some of the main dimensions of information that underlie the mobile telephony domain.

The proposed data mart presented below refers to mobile telephony consumer (residential) customers. Business accounts typically include multiple lines in a single

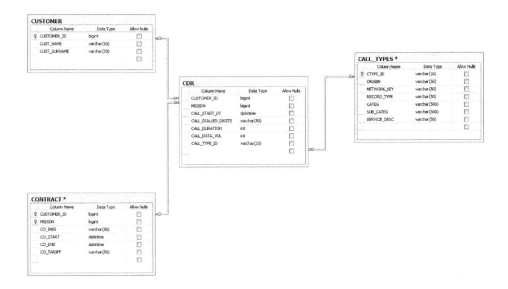

Figure 4.4   An example of a database diagram for a mobile telephony data mart.

corporate account and this fact should be considered in the design of the relevant data mart.

MOBILE TELEPHONY DATA AND CDRs

Each time a call is made its details are recorded and stored in CDRs. CDRs carry all the information required to build the usage profile of each customer, so they comprise the main data source for extracting and retrieving behavioral data. Although the format of the CDRs and the information logged vary among operators, typically such records contain information about the type of call (voice, SMS, etc.), the call originator (who made the call or A-number), the called party (who received the call or B-number), the call duration/traffic, the amount billed for each call, and so on.

CDRs are logged in a relevant table which typically contains information like that shown in Table 4.16.

Transforming CDR Data into Marketing Information

CDRs contain golden nuggets of information waiting to be refined. Their information is not directly usable. It requires data processing so that the underlying usage patterns are uncovered. Simple data management operations such as SQL queries or IBM SPSS Modeler data preparation commands can be applied to extract useful information and turn CDRs from simple transactional loggings into valuable customer information.

Table 4.16 An example of CDR information.

Field name	Field description
Record_ID	A sequence number identifying the record
A_Number	The number making the call – the MSISDN/SIM number
B_Number	The number receiving the call
Call_Start_Date	When the call started (date and time)
Call_Duration	The duration of the call
Data_Volume	What volume of data was transferred (using GPRS)
Call_Result	The result of the call (answered, busy, etc.)
Origination_ID	The identity of the source
Termination_ID	The identity of the destination
Call_Type	The type of call (voice, SMS, etc.)
Rated_Amount	The amount rated for the call
Billed_Amount	The amount actually charged for the call and to be billed

For example, the following SQL queries can identify if calls have been made during working days and working hours respectively:

```
case when datename(weekday,CALL_START_DATE) in
('Saturday','Sunday') then 'F' else 'T' end

case when datename(hour,CALL_START_DATE) between 8 and 17
then 'T' else 'F' end
```

The breakdown of calls according to the call hour (working vs. non-working) can be extracted by using the IBM SPSS Modeler Derive node as shown in Figure 4.5.

The following SQL query can be applied to extract the first layer of the "network" of each MSISDN (telephone number/SIM card), also referred to as the outgoing community. The outgoing community of each MSISDN is defined as the count of distinct telephone numbers (B-numbers) that have been called by the MSISDN:

```
Select MSISDN,count(distinct B_NUMBER) as
DISTINCT_CALLED_NUMBERS
From dbo.CDR
Group by MSISDN
```

Finally, the following aggregation summarizes data at an MSISDN level and on a monthly basis by calculating the total number and total duration of calls by call type (the month of call has already been extracted from the field denoting the date/time of call):

```
Select MSISDN,CALL_TYPE,YEAR,MONTH
count(*) as NUM_CALLS,sum(CALL_DURATION) as DUR_CALLS
from dbo.CDR
group by MSISDN,CALL_TYPE,YEAR,MONTH
```

The settings for extracting the same information by using the respective IBM SPSS Modeler command (node) are shown in Figure 4.6.

Although things are more complicated in real-world situations, the method of creating useful information from the CDRs is similar to the one presented above.

CURRENT INFORMATION

Current tables refer to all customers and lines, including ex-customers and closed lines. These tables contain the most recent updates of the relevant information. For example, a current table can be used to denote the most recent status information

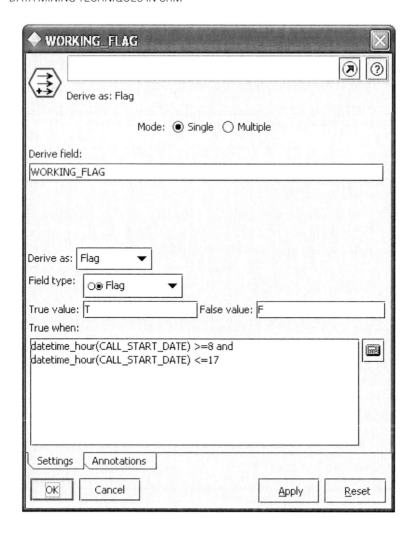

Figure 4.5 The IBM SPSS Modeler Derive node for breaking down calls according to the call hour (working vs. non-working).

of every line, covering both open and closed ones. In the case of a closed line, a date field indicates the closing date. This type of table can be used to record the latest customer demographics and the latest line details.

Customer Information

The first proposed table ("C_Customer"), Table 4.17, consolidates all basic customer information, including demographics and contact details. It is a table at

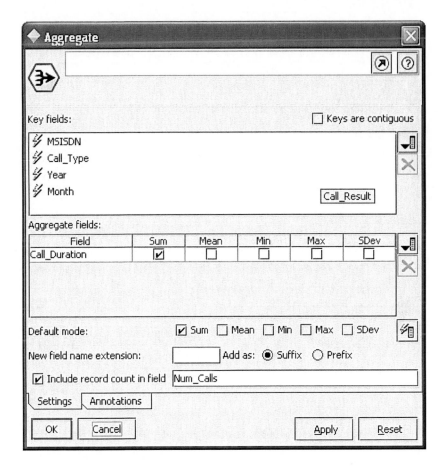

Figure 4.6   Grouping CDR data with the IBM SPSS Modeler Aggregate node.

the customer level, covering all customers that ever had a relationship with the network operator. The table's contents denote the most recent updates of the relevant information as well as the date of each record update.

Rate Plan History

The information in the rate plan table ("C_Rate_Plans"), Table 4.18, can be used to calculate the tenure of each customer and MSISDN, each MSISDN's current rate plan and status (open, closed), and the time of closing in the case of closed MSISDNs. Similarly, a list of recently churned (e.g., within the previous month) customers and MSISDNs can be easily produced.

Table 4.17 Basic customer information table.

Field name	Field type	Field description
Customer_ID	**Int**	**Customer identification number**
VAT_Num	Char	VAT number
ID_Num	**Char**	**ID number**
ID_Type_Code	Int	ID type – code number
Gender_Code	**Int**	**Gender – code number**
Marital_Status_Code	Int	Marital status – code number
Children_Num	**Int**	**Number of children**
House_Owner	Bit	House owner flag
Educational_Status_Code	**Int**	**Educational status – code number**
Occupation_Code	Int	Occupation – code number
Annual_Income_Code	**Int**	**Annual income band – code number**
Birth_Date	Datetime	Date of birth
Death_Date	**Datetime**	**Date of death, otherwise a control date**
Registration_Date	Datetime	First registration date
Closure_Date	**Datetime**	**Date of customer relationship closure**
First_Name	Char	First name
Last_Name	**Char**	**Last name**
Middle_Name	Char	Middle name
Home_Street	**Char**	**Home street name**
Home_Num	Char	Home street number
Home_City	**Char**	**Home city**
Home_State	Char	Home state
Home_ZIP	**Char**	**Home ZIP code**
Home_Country	Char	Home country
Work_Company	**Char**	**Name of company where customer works**
Work_Street	Char	Work street name
Work_Num	**Char**	**Work street number**
Work_City	Char	Work city
Work_State	**Char**	**Work state**
Work_ZIP	Char	Work ZIP code
Work_Country	**Char**	**Work country**
Home_Phone	Nchar	Home phone number
Work_Phone	**Nchar**	**Work phone number**
Home_Email	Char	Home e-mail
Work_Email	**Char**	**Work e-mail**
Last_Update	**Datetime**	**Last date of customer record update**

Customer_ID is the primary key (PK) of this table.

Table 4.18   The rate plan history table.

Field name	Field type	Field description
Customer_ID	Int	The customer ID
MSISDN	**Int**	**The MSISDN (mobile phone/SIM number)**
Rate_Plan	Int	The rate plan linked with the MSISDN
Paymethod	**Char**	**The most recent payment method (cash, standing order, credit card, etc.)**
Opening_Date	Datetime	The opening date of the rate plan
Closing_Date	**Datetime**	**The closing date of the rate plan (a future control date is used when the rate plan is still open)**
Opening_Type	Int	The opening type of the rate plan (e.g., new opening, transfer, portability, etc.)
Closing_Type	**Int**	**The closing type of the rate plan (e.g., voluntary churn, portability, transfer, involuntary churn, etc.)**
		If the rate plan is still open, the current status of the rate plan is recorded (e.g., active, barred, etc.)

Customer_ID, MSISDN, Rate_Plan and Opening_Date are the primary keys (PKs) of this table.

MONTHLY INFORMATION

The next proposed tables include monthly aggregations of key usage data. They can be used for summarizing and monitoring usage information such as traffic by call type, day, time zone, type of origin, and destination network. The availability of historical information allows customer usage to be tracked throughout the period covered by the data mart.

According to the requirements of each organization, a specific subset of key usage attributes could be monitored by weekly or even daily aggregations. For example, daily aggregations of the total number and duration of calls can be used for daily monitoring and reporting as well as for reactive campaigning. The daily information on top-ups (purchase of new credit) by prepaid customers, apart for reporting, can also be used for recency, frequency, monetary (RFM) monitoring in order to identify early churn signals.

Outgoing Usage

One of the key objectives of a data mart should be the identification of the usage patterns of each customer. The next proposed table ("M_Out_Usage"), Table 4.19,

Table 4.19   The outgoing usage table.

Field name	Field type	Field description
Year_Month	**Int**	**The number indicating the year and month of the data**
MSISDN	Int	The MSISDN (the telephone number that made the calls)
Service_Code	**Int**	**The service code that describes the type of call (voice, SMS, etc.)**
Network_Code	Int	The network code when calls were terminated (destination)
Roaming_Flag	**Bit**	**A flag indicating if the calls originated from a foreign country**
Peak_Flag	Bit	A flag indicating if the calls started during peak hours
Working_Flag	**Bit**	**A flag indicating if the calls started during working days**
Number_of_Calls	Int	Number of calls
Duration_of_Calls	**Real**	**Duration of calls (in seconds, for voice calls)**
Volume_of_Data	Real	Volume of data transferred (where applicable, for instance Internet calls)
Rated_Amount	**Real**	**The amount that was rated to the calls**
Billed_Amount	Real	The actual amount to be billed for the calls
Intercon_Amount	**Real**	**The interconnection cost of the calls**

Year_Month, MSISDN, Service_Code, Network_Code, Roaming_Flag, Peak_Flag, and Working_Flag are the primary keys (PKs) of this table.

contains monthly aggregations of outgoing traffic data. It records outgoing usage behavior over time, summarizing the number and duration/volume of outgoing calls by call type (voice, SMS, MMS, etc.), date/time, and type of destination network (calls to fixed lines, international calls, calls to another mobile network operator, etc.). This table is supplemented by a set of reference (lookup) tables that decode the call characteristics. Traffic is summarized at an MSISDN level. Further aggregations, for instance at a customer level, are also possible.

Top-up Information

Unlike postpaid customers who have a contract agreement and a billing arrangement with the network operator, prepaid customers purchase services in advance through top-ups. They can use their phone until they run out of credit. The addition of more credit to their account is referred to as topping up the account. Therefore, another data dimension, concerning only the prepaid customers, that should be tracked is the top-ups. The minimum information that should be monitored on a monthly basis includes the frequency (number) and value (credit amount added) by top-up type. Additionally, an organization might also follow the top-up recency (time since last top-up) and the trend of average times between top-ups, although the latter would require a detailed recording of each top-up.

Incoming Usage

The incoming usage table ("M_Inc_Usage"), Table 4.20, applies the logic of monthly aggregations presented above to the incoming traffic data. Incoming usage behavior is tracked over time with monthly aggregations that summarize the

Table 4.20   The incoming usage table.

Field name	Field type	Field description
Year_Month	**Int**	**The number that indicates the year and month of data**
MSISDN	Int	The MSISDN (the telephone number that received the calls)
Service_Code	**Int**	**The service code that describes the type of call (voice, SMS, etc.)**
Network_Code	Int	The network code from where the calls originated (source)
Roaming_Flag	**Bit**	**A flag indicating if the calls were received in a foreign country**
Peak_Flag	Bit	A flag indicating if the calls started during peak hours
Working_Flag	**Bit**	**A flag indicating if the calls started during working days**
Number_of_Calls	Int	Number of calls
Duration_of_Calls	**Real**	**Duration of calls (in seconds, for voice calls)**
Volume_of_Data	Real	Volume of data transferred (where applicable, for instance Internet calls)
Roaming_Amount	**Real**	**The amount that was rated to the calls due to incoming roaming**
Intercon_Amount	Real	The incoming amount charged to third parties for calls made to the specific MSISDN

Year_Month, MSISDN, Service_Code, Network_Code, Roaming_Flag, Peak_Flag, and Working_Flag are the primary keys (PKs) of this table.

Table 4.21   The outgoing network table.

Field name	Field type	Field description
Year_Month	**Int**	**The number that indicates the year and month of data**
MSISDN	Int	The MSISDN (the number that made the calls)
Called_Num	**Int**	**The B-number (the number that received the calls)**
Network_Code	Int	The network code that calls were terminated (destination)
Number_of_Calls	**Int**	**Number of calls**
Duration_of_Calls	Real	Duration of calls (in seconds, for voice calls)

Year_Month, MSISDN, Network_Code, and Called_Number are the primary keys (PKs) of this table.

number and duration/volume of incoming calls by call type, date/time, and type of source network.

Outgoing Network

The proposed outgoing network table ("M_Out_Network"), Table 4.21, can be used to extract the outgoing community size of each customer, which is the number of distinct telephone numbers called. The community size can also be examined by network destination, revealing the on-net and the off-net community of each MSISDN. The on-net community refers to internal calls made to other MSISDNs of the same network. The off-net community refers to calls made to external networks and can be further broken down to reveal the community belonging to major competitors.

Incoming Network

The proposed incoming network table ("M_Inc_Network"), Table 4.22, deals with incoming traffic and indicates the sources (networks and telephone numbers) of received calls.

LOOKUP INFORMATION

The main use of the lookup tables in the case of the mobile telephony mining data mart is to decode and describe the different types of calls (voice, SMS, etc.), rate plans (the type of contract that the customer has with the operator), and network types (international, fixed line destination, internal network, external network,

Table 4.22   The incoming network table.

Field name	Field type	Field description
Year_Month	**Int**	**The number that indicates the year and month of data**
MSISDN	Int	The MSISDN (the number that received the calls)
Caller_Num	**Int**	**The number that called the MSISDN**
Network_Code	Int	The network code from where the calls originated
Number_of_Calls	**Int**	**Number of calls**
Duration_of_Calls	Real	Duration of calls (in seconds, for voice calls)

Year_Month, MSISDN, Network_Code, and Caller_Number are the primary keys (PKs) of this table.

Table 4.23   A lookup table for rate plan codes.

Field name	Field type	Field description
Rate_Plan_Code	**Int**	**The rate plan code**
Rate_Plan_Desc	Char	The rate plan description
Rate_Plan_Group	**Char**	**The rate plan group**
Monthly_Fee	Int	The monthly fee of the rate plan
Free_Minutes	**Int**	**The free minutes offered by the rate plan**
Free_SMS	Int	The free SMSs offered by the rate plan
Free_MMS	**Int**	**The free MMSs offered by the rate plan**
Free_Data	Int	The free volume of data offered by the rate plan

Rate_Plan_Code is the primary key (PK) of this table.

competitor, etc.). Quite often the call and destination type are handled together with a single code also used for rating – billing. The lookup tables presented below complement the proposed data mart tables presented in the previous sections.

Rate Plans

The rate plan table ("L_Rate_Plan"), Table 4.23, provides a description of the rate plan codes offered by the network operator. Information specific for each rate plan (monthly fee, free airtime, etc.) should also be included in this table. The table should also enable the categorization of customers as prepaid or postpaid, according to their rate plan group.

Service Types

Usage behavior by service (call) type is monitored through a respective service code field included in the usage tables presented above. Now a lookup table (Table 4.24), decodes this information into understandable call types.

Table 4.24 A lookup table for service types.

Field name	Field type	Field description
Service_Code	Int	The service code of the call
Service_Desc	Char	The service description of the call

Service_Code is the primary key (PK) of this table.

Table 4.25 A typical list of mobile telephony service types.

Service types
Voice calls
SMS calls
MMS calls
Internet calls
WAP calls
Video calls
Events/service calls
Mailbox calls
Other

Typical service types are given in Table 4.25.

Networks

The last proposed lookup table ("L_Network table"), Table 4.26, decodes the different types of source (for incoming calls) and destination (for outgoing calls) networks.

A typical categorization of network types is presented in Table 4.27.

THE MCIF FOR MOBILE TELEPHONY

Even if the data mart covered the majority of mining and marketing needs, its information would typically require additional data preparation before being appropriate for use in data mining applications. This is where the MCIF fits in.

Table 4.26   A lookup table for network codes.

Field name	Field type	Field description
Network_Code	**Int**	**The network code where the call was terminated/originated**
Network_Desc	Char	The description of the network code where the call was terminated/originated
Network_Group	**Char**	**The network group (see following categorization, first level)**
Network_Subgroup	Char	The network subgroup (see following categorization, second level)

Network_Code is the primary key (PK) of this table.

Table 4.27   A typical categorization of mobile telephony network types.

Network group	Network subgroup
Fixed	• Fixed national
International	• International
On-net calls	• Internal network numbers • Customer service • Call center
Off-net calls	• Mobile competitor 1 • Mobile competitor 2 • ⋮ • Mobile competitor n
Information lines	• Information lines

The MCIF is a flat table which combines the most important blocks of information to provide an integrated view of the customer's relationship with the organization. The data mart is a general purpose data repository, a transitional stage through which operational data pass before being used for analysis. The MCIF is the last step in the data preparation procedure, with data that can immediately (or with a little extra processing) support the majority of the data mining applications. The MCIF presented here is at an MSISDN level, although the data mart design also allows the MCIF to be built at a different level (e.g., at a customer level) if required.

Although the contents of the MCIF should take into account the specific mining needs and plans of each organization, we present in Table 4.28 a listing of fields in an attempt to delineate the data dimensions typically required for identifying the traffic patterns and usage profile of mobile phone customers.

Table 4.28 An indicative listing of the contents of the MCIF for mobile telephony.

Field name	Field description
Customer socio-demographics and line information:	
CUSTOMER_ID	Unique customer (account) identifier
MSISDN	The mobile phone number
CONTRACT_ID	The contract ID
VAT_NUM	Value added tax number
GENDER	Gender (male, female, n/a)
DATE_OF_BIRTH	Date of birth
OCCUPATION_CODE	Occupation description (e.g., doctor, lawyer, n/a, etc.)
HOME_ZIP	ZIP code
REGISTRATION_DATE	Date of registration
TENURE	Time (days) since registration
PAYMETHOD	Payment method (cash, standing order, credit card, etc.)
RATE_PLAN	Rate plan
CUSTOMER_TYPE	Residential postpaid/contractual customers, residential prepaid customers
STATUS	Status of MSISDN (active, disconnected, barred, etc.)
Community:	
OUT_COMMUNITY_TOTAL	Total outgoing community: number of distinct phone numbers that the holder called
OUT_COMMUNITY_ONNET	Outgoing on-net community (same network – internal phone numbers)
OUT_COMMUNITY_OFFNET	Outgoing off-net community (other network – external phone numbers)
IN_COMMUNITY_TOTAL	Total incoming community
IN_COMMUNITY_ONNET	Incoming on-net community (same network – internal phone numbers)
IN_COMMUNITY_OFFNET	Incoming off-net community (other network – external phone numbers)
Number of calls by service type:	
VOICE_OUT_CALLS	Number of outgoing voice calls
SMS_OUT_CALLS	Number of outgoing SMS calls
MMS_OUT_CALLS	Number of outgoing MMS calls
EVENTS_CALLS	Number of event calls
INTERNET_CALLS	Number of Internet calls
MAILBOX_CALLS	Number of mailbox calls
CS_CALLS	Number of customer service calls

Table 4.28 (*continued*)

Field name	Field description
WAP_CALLS	Number of WAP calls
TOTAL_OUT_CALLS	Total number of outgoing calls (includes all call types)
DAYS_OUT	Number of days with outgoing usage
VOICE_IN_CALLS	Number of incoming voice calls
SMS_IN_CALLS	Number of incoming SMS calls
MMS_IN_CALLS	Number of incoming MMS calls
TOTAL_IN_CALLS	Total number of incoming calls (includes all call types)
DAYS_IN	Number of days with incoming usage
Minutes/traffic by service type:	
VOICE_OUT_MINS	Total number of minutes of outgoing voice calls
EVENTS_TRAFFIC	Total events traffic
MMS_OUT_VOLUME	Total outgoing MMS volume
GPRS_TRAFFIC	Total GPRS traffic
VOICE_IN_MINS	Total number of minutes of incoming voice calls
ACD_OUT	Outgoing average call duration (minutes/outgoing calls)
ACD_IN	Incoming average call duration (minutes/outgoing calls)
Usage by destination/source network:	
OUT_CALLS_ONNET	Number of outgoing calls on-net (internal network)
OUT_CALLS_A	Number of outgoing calls to mobile competitor A
OUT_CALLS_B	Number of outgoing calls to mobile competitor B
OUT_CALLS_C	Number of outgoing calls to mobile competitor C
OUT_CALLS_FIXED	Number of outgoing calls to fixed networks
OUT_CALLS_INTERNATIONAL	Number of outgoing calls to international networks
OUT_CALLS_ROAMING	Number of outgoing roaming calls (voice calls made in a foreign country)
OUT_MINS_ONNET	Total number of outgoing on-net minutes
OUT_MINS_A	Total number of outgoing minutes to competitor A
OUT_MINS_B	Total number of outgoing minutes to competitor B
OUT_MINS_C	Total number of outgoing minutes to competitor C
OUT_MINS_INTERNATIONAL	Total number of outgoing minutes to international networks
OUT_MINS_FIXED	Total number of outgoing minutes to fixed networks
OUT_MINS_ROAMING	Total number of outgoing roaming minutes
IN_CALLS_ONNET	Number of incoming calls on-net (internal network)

(*continued overleaf*)

Table 4.28 *(continued)*

Field name	Field description
IN_CALLS_A	Number of incoming calls from mobile competitor A
IN_CALLS_B	Number of incoming calls from mobile competitor B
IN_CALLS_C	Number of incoming calls from mobile competitor C
IN_CALLS_FIXED	Number of incoming calls from fixed networks
IN_CALLS_INTERNATIONAL	Number of incoming calls from international networks
IN_CALLS_ROAMING	Number of incoming roaming calls (voice calls received in a foreign country)
IN_MINS_ONNET	Total number of incoming on-net minutes
IN_MINS_A	Total number of incoming minutes from competitor A
IN_MINS_B	Total number of incoming minutes from competitor B
IN_MINS_C	Total number of incoming minutes from competitor C
IN_MINS_INTERNATIONAL	Total number of incoming minutes from international networks
IN_MINS_FIXED	Total number of incoming minutes from fixed networks
IN_MINS_ROAMING	Total number of incoming roaming minutes
Usage by day/hour:	
OUT_CALLS_PEAK	Number of outgoing calls in peak hours
OUT_CALLS_OFFPEAK	Number of outgoing calls in off-peak hours
OUT_CALLS_WORK	Number of outgoing calls in work days
OUT_CALLS_NONWORK	Number of outgoing calls in non-work days
IN_CALLS_PEAK	Number of incoming calls in peak hours
IN_CALLS_OFFPEAK	Number of incoming calls in off-peak hours
IN_CALLS_WORK	Number of incoming calls in work days
IN_CALLS_NONWORK	Number of incoming calls in non-work days
Revenue:	
OUTGOING_REVENUE	Total amount of revenue from outgoing usage
INCOMING_REVENUE	Total amount of revenue from incoming usage (inter-connection costs paid by other operators to recipient's operator)
MARPU	Total MARPU (Marginal Average Revenue Per User) value index

---

Examples of Derived Measures Used to Enrich the MCIF

An indicative list of more complex KPI examples for mobile telephony is as follows:

- Percentage (%) of peak calls: outgoing peak calls as a percentage of total outgoing calls.
- Percentage (%) of working day calls.
- Percentage (%) of SMS calls: outgoing SMS calls as a percentage of total outgoing calls.
- Percentage (%) of on-net outgoing community.
- Percentage (%) of the total number of calls made to the three most frequently called numbers.
- Ratio of total outgoing to total incoming calls.
- Percentage (%) of outgoing voice calls with a short duration (< 60 seconds).
- Number of distinct months with roaming calls.
- Outgoing usage delta: compares the total outgoing calls (or minutes) of the most recent period, for example, the last three months, to those of the previous three months and denotes the respective percentage difference.
- Flag of Internet usage.
- Coefficient of variation of outgoing calls: a measure of the weekly variation of outgoing calls, defined as the standard deviation of the weekly number of outgoing calls divided by the weekly mean. It requires weekly aggregates of call data.
- Number of different service (call) types.
- Number of rate plan changes (divided by the respective tenure).

THE MINING DATA MART FOR RETAILERS

As in any other industry, retailers need to continuously monitor the relationship with their customers. Fortunately, all customer transactions are recorded in detail at the point of sale (POS), logging information on:

- **What:** The detailed universal product code (UPC) of the items purchased.
- **When:** Time/day.

- **Where:** The specific store/department/channel concerned.
- **How:** The payment type, for example, cash, credit card, and so on, of each purchase.

Data obtained from the point of sale, when appropriately mined, can lead to a clear identification of customer characteristics. A specific problem that retailers have to tackle in order to track the transaction "trail" of each customer is the "personalization" of the purchase data. A form of identification is required to link transactional data to the customers. This identification is usually made possible through a loyalty scheme. A loyalty scheme's main purpose is to tighten the relationship with the customers and to promote loyalty through rewards and bonus points. Furthermore, it also provides a means of collecting customer information and of relating the transactions to the customers.

The scope of a mining data mart for retailers is to provide an integrated view of all customer attributes, allowing customer behavior to be monitored over time.

A key concept in retailing data is the "transaction." Hereafter the term transaction will refer to each purchase event. A transaction can be thought of as a purchase basket or a discrete order and typically involves the purchase of multiple product items.

The information contained in the retailing data mart should capture all transactions and all customer-related information, covering, at a minimum, the following data dimensions:

- Registration data, collected during the loyalty scheme application process.
- RFM attributes, overall or for each product category. More specifically:

  - The time since the last transaction (purchase event) or since the most recent visit day.
  - The frequency (number) of transactions or number of days with purchases.
  - The value of total purchases.

- Relative spending in each product category/subcategory.
- Preferred stores/departments.
- Time and day of customer visits, also including special occasions (summer, Christmas, etc.) and sales periods.
- Payment (cash, credit card, etc.) type and channel of purchase (store, phone, Internet).
- Campaign contacts and responses.
- Use of vouchers and reward points.
- Recorded complaints and results.

A period of two years is proposed as the time frame for the data mart in order to support the majority of the data mining applications, including predictive modeling. Once again the main information tables should be combined with descriptive lookup tables which can facilitate further categorizations of the data dimensions. A lookup table of critical importance is the one used to describe and organize the product codes (UPCs) into a multilevel hierarchy. This hierarchy maps each product into corresponding categories and subcategories, enabling the grouping of purchases at the desired product level. In a dynamic retailing environment, the products offered change continuously. Therefore the product categorization table should be well organized and the product taxonomy should be frequently updated to accurately account for all such product changes.

TRANSACTION RECORDS

The data collected at the point of sale log all the details of each transaction, carrying information about the type of products purchased, the value and amount of each product bought, the store visited, and the exact date and time when the transaction took place. Although the format and the exact contents of the transaction records vary among retailers, a typical example of the information logged is given in Table 4.29.

The above data form the basic information blocks for building the retailing data mart proposed in the next section. The scope of this data mart is to provide a good starting point for the majority of the analytical tasks, without having to undertake rigorous data management of raw transactional records. For extra information not covered by the data mart, data miners can always return to the raw transactional data to make use of their detailed information.

CURRENT INFORMATION

Current tables contain information about current and past customers and only carry the latest update of the relevant information. A typical example of a current table is the one used to store the latest socio-demographic information of all registered customers.

Customer Information

Apart from transactional data, the data mart should incorporate all the customer socio-demographic and contact information. Information related to the customer's life-stage, such as age and marital status, may determine at a certain level his or

Table 4.29 An example of point-of-sale transactional data.

Field name	Field description
Transaction_Line_ID	**An identifier assigned to each record of data (line)**
Transaction_ID	The transaction ID uniquely identifies each transaction. Each transaction is a unique purchase event comprising many records of data (Transaction_Line_ID) and many purchased items
Transaction_Type	**The type of transaction (store, phone order, Internet order)**
Customer_ID	The loyalty card number presented by the customer
Date_and_Time	**The timestamp of the raw transaction start time**
Store_Code	The store code of transaction
Check_Out_Code	**The checkout code of transaction**
Item_Code	The UPC of the purchased item (see next "L_Items" lookup table)
Num_of_Items	**The number of purchased items**
Rated_Amount	The amount rated for the purchased item
Discount_Amount	**The amount discounted from the rated amount**
Payment_Type	**The payment type (cash, credit card, voucher)**

her purchasing habits, therefore it should not be omitted from the analytical input inventory. This type of information is typically collected from the application for a loyalty card. One issue to bear in mind about this kind of information is its validity. Collected information may be incomplete, incorrect, or outdated, therefore it should be used cautiously for data mining purposes. This type of information, if possible, should be checked and updated periodically.

The first table for the proposed data mart is a current table covering the customer's socio-demographic and contact information, Table 4.30.

MONTHLY INFORMATION

These tables include monthly aggregations of key purchase information, such as the number and volume of transactions. Data are summarized at a customer level.

Transactions

The first monthly table ("M_Transactions"), Table 4.31, summarizes the frequency (number) of transactions at a customer level, by store/channel, day of the week, and time zone.

Table 4.30   Basic customer information table.

Field name	Field type	Field description
Customer_ID	**Int**	**Customer ID or loyalty card ID**
VAT_Num	Char	VAT number
Gender_Code	**Int**	**Gender – code number**
Marital_Status_Code	Int	Marital status – code number
Children_Num	**Int**	**Number of children**
Child_Birth_Date	Datetime	The date of birth of first child
Educational_Status_Code	**Int**	**Educational status – code number**
Occupation_Code	Int	Occupation – code number
Annual_Income_Code	**Int**	**Annual income band – code number**
Birth_Date	Datetime	Date of birth
Death_Date	**Datetime**	**Date of death, otherwise a control date**
Registration_Date	Datetime	First registration date
Closure_Date	**Datetime**	**Date of customer relationship closure**
First_Name	Char	First name
Last_Name	**Char**	**Last name**
Middle_Name	Char	Middle name
Home_Street	**Char**	**Home street name**
Home_Num	Char	Home street number
Home_City	**Char**	**Home city**
Home_State	Char	Home state
Home_ZIP	**Char**	**Home ZIP code**
Home_Country	Char	Home country
Fixed_Phone	**Nchar**	**Home phone number**
Mobile_Phone	Nchar	Work phone number
Home_Email	**Char**	**Home e-mail**
Work_Email	Char	Work e-mail
Last_Update	**Datetime**	**Last date of customer record update**

Customer_ID is the primary key (PK) of this table.

Table 4.31 The transactions table.

Field name	Field type	Field description
Year_Month	**Int**	**The number that indicates the year and month of data**
Customer_ID	Int	Customer ID or loyalty card ID
Store_Code	**Int**	**The code of the store where the transaction took place (also includes the channel information, in case of an Internet or phone order)**
Morning_Transactions	Int	Number of transactions made from 08:00 to 17:00
Weekend_Transactions	**Int**	**Number of transactions on Saturdays and Sundays**
Num_of_Transactions	Int	Total number of transactions/visits
Num_of_Items	**Int**	**Total number of items purchased**
Total_Paid_Amount	Real	Total amount paid
Date_Last_Transaction	**Datetime**	**The date and time of the last (most recent) transaction within the specific month**

Year_Month, Customer_ID, and Store_Code are the primary keys (PKs) of this table.

Monitoring Key Usage Attributes on a More Frequent Basis

According to the requirements of each organization, a specific subset of key usage attributes could be monitored by weekly or even daily aggregations.

Especially for retailers, the daily purchases should be monitored carefully in order to investigate seasonal and market event effects. A "diary" lookup table can be used to map dates with special occasions, such as sales seasons, holidays, weekends, and so on.

Purchases by Product Groups

The next monthly table ("M_Purchases"), Table 4.32, records the relative spending of each customer by product groups. The product level selected for this aggregated

Table 4.32 Purchases by product groups.

Field name	Field type	Field description
Year_Month	**Int**	**The number that indicates the year and month of data**
Customer_ID	Int	Customer ID or loyalty card ID
Item_Group	**Int**	**UPC group: the highest level of the product hierarchy for which summarized data are kept**
Num_of_Items	Int	Total number of items purchased
Total_Rated_Amount	**Real**	**The total rated amount**
Total_Discount_Amount	Real	The total discount amount

Year_Month, Customer_ID, and Item_Group are the primary keys (PKs) of this table.

table should correspond to the highest hierarchy level that can address the most frequent analytical needs. Nevertheless, as mentioned above, apart from pre-aggregated purchase data, we can always revisit the raw transactional data and the relevant lookup table to extract information at a lower product level.

LOOKUP INFORMATION

A well-organized lookup table is necessary to classify the product codes in a multilevel product hierarchy.

The Product Hierarchy

The next table ("L_Items"), Table 4.33, defines a multilevel product hierarchy that can be used to describe and categorize the items into product groups and subgroups according to their type. The product hierarchy lookup table can be combined with the point-of-sale transaction records to enable summarization at the selected product level, providing more detailed information than the aggregated monthly tables. The hierarchy level to be used depends on the specific needs of the organization. Additional summarizations, for example, by brand (brand labels, private labels, etc.), are also feasible.

An indicative list of product groups at a high level in the product hierarchy is presented in Table 4.34.

Table 4.33 A lookup table for the product hierarchy.

Field name	Field type	Field description
Item_Code	**Int**	**The UPC of the item**
Item_Description	Char	The detailed item description
Item_Brand	**Char**	**The manufacturer of the specific item**
Item_Group	Char	The hierarchy group of the item
Item_SubGroup	**Char**	**The hierarchy subgroup of the item**

Item_Code is the primary key (PK) of this table.

Table 4.34 An indicative list of product groups at a high level in the product hierarchy.

Product groups
Apparel/shoes/jewelry
Baby
Electronics
Computers
Food and wine
Health and beauty
Pharmacy
Sports and outdoors
Books/press
Music, movies, and videogames
Toys
Home
Other

THE MCIF FOR RETAILERS

The MCIF for retailers should include all the information required to identify the purchasing habits of customers. Although its contents depend on the particular organization and they should reflect its specific marketing and analytical objectives, Table 4.35 presents an indicative listing of the main data dimensions typically used for data mining purposes.

Table 4.35   An indicative listing of the contents of the MCIF for retailers.

Field name	Field description
Customer socio-demographics:	
CUSTOMER_ID	Customer ID or loyalty card ID
GENDER	Gender of the registered customer
MARITAL_STATUS	Marital status of the registered customer
CHILDREN_NUM	Number of children
CHILD_BIRTH_DATE	Date of birth of first child
EDUCATIONAL_STATUS	Educational status of registered customer
OCCUPATION	Occupation of registered customer
ANNUAL_INCOME	Annual income band
BIRTH_DATE	Date of birth
REGISTRATION_DATE	Loyalty card registration date
Customer transactions (visits):	
LAST_VISIT_DATE	Date of last visit to all stores
NUM_TRANSACTIONS	Total number of transactions at all stores
DIFFERENT_STORES	Total number of different stores visited
MORNING_TRANSACTIONS	Total number of transactions from 08:00 to 17:00
WEEKEND_TRANSACTIONS	Total number of transactions on Saturdays and Sundays
NUM_PHONE_ORDERS	Total number of transaction orders by phone
NUM_INTERNET_ORDERS	Total number of transaction orders over the Internet
Customer purchases:	
NUM_OF_ITEMS	Total number of items purchased
AMOUNT_PAID	Total amount paid for purchases
AMOUNT_DISCOUNT	Total amount discounted

(continued overleaf)

Table 4.35   (*continued*)

Field name	Field description
AMOUNT_CASH	Total purchase amount paid with cash
AMOUNT_CREDIT_CARDS	Total purchase amount paid with credit cards
AMOUNT_VOUCHERS	Total purchase amount paid with vouchers
AMOUNT_POINTS	Total amount of reward points redeemed
RFM	Recency frequency monetary index
TOTAL_REVENUE	Total revenue
Purchases by product groups:[a]	
AMOUNT_BRAND_LABELS	Total amount paid for brand labels
AMOUNT_PRIVATE_LABELS	Total amount paid for private labels
AMOUNT_APPAREL_SHOES_ JEWELRY	Total amount paid for apparel, shoes, and jewelry
AMOUNT_BABY	Total amount paid for baby products
AMOUNT_ELECTRONICS	Total amount paid for electronics
AMOUNT_COMPUTERS	Total amount paid for computers
AMOUNT_FOOD_WINE	Total amount paid for food and wine
AMOUNT_HEALTH_BEAUTY	Total amount paid for health and beauty products
AMOUNT_PHARMACY	Total amount paid for pharmaceutical products
AMOUNT_SPORTS_OUTDOORS	Total amount paid for sports and outdoors products
AMOUNT_BOOKS_PRESS	Total amount paid for books and press (newspapers and magazines)
AMOUNT_MUSIC_MOVIES_ VIDEOGAMES	Total amount paid for music, movies, and videogames
AMOUNT_TOYS	Total amount paid for toys
AMOUNT_HOME	Total amount paid for home products
AMOUNT_OTHER	Total amount paid for other products

[a]The following is an indicative breakdown of purchases by product groups (the product groups and the level of hierarchy selected for the grouping depend on the retailer).

Examples of Derived Measures Used to Enrich the MCIF

An indicative list of more complex KPI examples for retailers is as follows:

- Basket size: average transaction amount and average number of items per transaction.
- Percentage (%) of purchase amount spent on private labels.
- Percentage (%) of purchase amount spent per product group/subgroup.
- Monthly average number of visits within the period examined.
- Total number of distinct product groups/subgroups purchased.
- Average time interval between two sequential visits.
- Flag of order over the Internet.

SUMMARY

Data mining can play a significant role in the success of an organization, but only if it is integrated into the organization's operations. Data mining models need data as input and produce valuable results as output. A standardized procedure is required for the deployment of the derived results as well as for obtaining the input data.

In this chapter we have dealt with the issue of input data and suggested the main data dimensions typically required to cover the analytical needs of the banking, retailing, and mobile telephony industries. We also proposed a way to organize the data for the needs of data mining applications. Customer attributes assessed as necessary for marketing and mining purposes should be extracted from various data sources, transformed, and loaded into a central mining repository, the mining data mart, which should be designed to cover the majority of the analytical needs of the organization. The mining data mart should enable analysts to monitor the key customer attributes over time, without delays for ad hoc data acquisition and preparation, and without additional requests for implementations from IT. The contents of the data mart are the main information blocks used to build the MCIF, a flat table, typically at a customer level, which integrates the information of each customer into a single record, a customer "signature" record. The MCIF's purpose is to be an almost-ready input file for most of the upcoming analytical applications.

CHAPTER FIVE

Customer Segmentation

AN INTRODUCTION TO CUSTOMER SEGMENTATION

Customer segmentation is the process of dividing customers into distinct, meaningful, and homogeneous subgroups based on various attributes and characteristics. It is used as a differentiation marketing tool. It enables organizations to understand their customers and build differentiated strategies, tailored to their characteristics.

Traditionally organizations, regardless of the industry they operate in, tend to use market segmentation schemes that are based on demographics and value information. Over the past few decades organizations have been deciding their marketing activities and developing their new products and services based on these simple, business rule segments. In today's competitive markets, this approach is not sufficient. On the contrary, organizations need to have a complete view of their customers in order to gain a competitive advantage. They also need to focus on their customers' needs, wants, attitudes, behaviors, preferences, and perceptions, and to analyze relevant data to identify the underlying segments. The identification of groups with unique characteristics will enable the organization to manage and target them more effectively with, among other things, customized product offerings and promotions.

There are various segmentation types according to the segmentation criteria used. Specifically, customers can be segmented according to their value, socio-demographic and life-stage information, and their behavioral, need/attitudinal, and loyalty characteristics. The type of segmentation used depends on the specific business objective. In the following sections we will briefly introduce all the segmentation types before focusing on the behavioral and value-based segmentations that are the main topics of this book.

Data Mining Techniques in CRM: Inside Customer Segmentation   K. Tsiptsis and A. Chorianopoulos
© 2009 John Wiley & Sons, Ltd

In general, the application of a cluster model is required to reveal the segments, particularly if we need to combine a large number of segmentation attributes. As opposed to business rules, a cluster model is able to manage a large number of attributes and reveal data-driven segments which are not known in advance.

Segments are created by analyzing the customer information residing in the organization's data marts and data warehouse. For specific segmentation types, a market survey is required to collect the necessary segmentation criteria since the relevant attributes are unavailable (needs/attitudes, loyalty information) or partially available or outdated (demographics) in the organization's data systems.

A market survey is also needed in order to analyze the entire market, including customers of competitor organizations. In the framework of this book, though, we are going to focus on segmenting the existing customer base.

Consumers and business customers have inherent differences, as summarized in the next section. Due to these differences, the two markets require a different segmentation approach and the use of different segmentation criteria. In this book we are going to focus mostly on segmentation for consumer markets.

SEGMENTATION IN MARKETING

Segmentation identifies the different customer typologies, facilitating the development of differentiated marketing strategies to better serve and manage the customers.

Segmentation in marketing can be used for the following:

- **Greater understanding of customers** to support the identification of new marketing opportunities.
- **Design and development of new products/services and product bundles tailored to each segment's characteristics** rather than the mass market.
- **Design of customized product offering strategies to existing customers** according to each segment's identified wants and needs.
- **Offering tailored rewards and incentives.**
- **Selecting the appropriate advertising and communication message and channel.**
- **Selecting the appropriate sales and service channel.**
- **Determining the brand image and the key product benefits to be communicated** based on the specific characteristics of each segment.
- **Differentiation in customer service** according to each segment's importance.
- **More effective resource allocation** according to the potential return from each segment.
- **Prioritization of the marketing initiatives** which aim at customer retention and development according to each segment's importance and value.

A successful segmentation scheme:

- Addresses the business objective set by the organization.
- Identifies clearly differentiated segments with unique characteristics and needs.
- Accurately assigns each individual customer to a segment.
- Provides the opportunity for profit in the identified segments.
- Comprises "identifiable" segments with recognizable profiles and characteristics.
- Is actionable and applicable to business operations. The organization can design effective marketing actions for the revealed segments.

SEGMENTATION TASKS AND CRITERIA

There is no magic segmentation solution that adequately covers all business situations. Different criteria and segmentation methods are appropriate for different situations and business objectives.

Segmentation is used in strategic marketing to support multiple business tasks. The starting point should always be the particular business situation. Business objectives and success criteria should be clearly defined before actually starting the analysis. The appropriate segmentation criteria and the required analytical tools should then be selected.

Table 5.1 lists the appropriate segmentation types and the required tools and techniques per business task.

SEGMENTATION TYPES IN CONSUMER MARKETS

There are various criteria for customer segmentation that can be used to optimize consumer marketing. As mentioned above, different segmentation types are used for different business situations. The following segmentation types are most widely used:

1. **Value based:** In value-based segmentation customers are grouped according to their value. This is one of the most important segmentation types since it can be used to identify the most valuable customers and to track value and value changes over time. It is also used to differentiate the service delivery strategies and to optimize the allocation of resources in marketing initiatives.
2. **Behavioral:** This is a very efficient and useful segmentation type. It is also widely used since it presents minimal difficulties in terms of data availability. The required data include product ownership and utilization data which are

Table 5.1   Segmentation types and business tasks.

Business situation/task	Appropriate segmentation criteria	Analytical tools and techniques
New product design and development	Needs/attitudinal and behavioral	Combination of data mining and market survey/factor and cluster analysis
Design of customized product offering strategies	Behavioral	Data mining/factor and cluster analysis
Brand image and key product benefits to be communicated	Needs/attitudinal	Market surveys/factor analysis and cluster analysis
Differentiated customer service	Customer value in combination with other attributes, for example, age (tenure) of customer	Binning (grouping in tiles) of customers according to their value (e.g., low $n\%$, medium $n\%$, top $n\%$) and cross-tabulating with other attributes, for example, value and oldness of customer
Resource allocation and prioritization of the marketing interventions that aim at customer development and retention	Customer value supplemented by deep understanding of what drives customer decision to buy and/or to churn	Value tiles and market survey to identify drivers of decisions to buy and/or to churn
Identifying target groups for campaigns	Propensity scores derived from relevant classification models	Data mining using classification modeling – grouping customers according to their propensity scores and their likelihood to churn and/or to buy

usually stored and available in the organization's databases. Customers are divided according to their identified behavioral and usage patterns. This type of segmentation is typically used to develop customized product offering strategies. Also, for new product development, and the design of loyalty schemes.

3. **Propensity based:** In propensity-based segmentation customers are grouped according to propensity scores, such as churn scores, cross-selling scores, and

so on, which are estimated by respective classification (propensity) models. Propensity scores can also be combined with other segmentation schemes to better target marketing actions. For instance, the value-at-risk segmentation scheme is developed by combining propensities with value segments to prioritize retention actions.

4. **Loyalty based:** Loyalty segmentation involves the investigation of the customers' loyalty status and the identification of loyalty-based segments such as loyals and switchers/migrators. Retention actions can then be focused on high-value customers with a disloyal profile whereas cross-selling on prospectively loyal customers.

5. **Socio-demographic and life-stage:** This type reveals different customer groupings based on socio-demographic and/or life-stage information such as age, income, marital status. This type of segmentation is appropriate for promoting specific life-stage-based products as well as supporting life-stage marketing.

6. **Needs/attitudinal:** This segmentation type is typically based on market research data and identifies customer segments according to their needs, wants, attitudes, preferences, and perceptions pertaining to the company's services and products. It can be used to support new product development and to determine the brand image and key product features to be communicated.

VALUE-BASED SEGMENTATION

Value-based segmentation is the process of dividing the customer base according to value. It should be emphasized that this is not a one-off task. It is vital for the organization to be able to track value changes across time. The organization should monitor and, if possible, intervene in order to prevent downward and encourage upward migrations.

A prerequisite for this segmentation scheme is the development of an accurate and credible procedure for determining the value of each customer, on a periodic basis, preferably at least monthly, using day-to-day inputs on revenues and costs. Value-based segmentation is developed through simple computations and does not involve the application of a data mining model. Specifically, the identification of the value segments involves sorting customers according to their value and their binning in chunks of equal size, for example, of 10% named quantiles. These quantiles are the basis for the development of value segments of the form low $n\%$, medium $n\%$, top $n\%$. A list of typical value segments is as follows:

- **Gold:** Top 20% of customers with the highest value.
- **Silver:** 30% of customers with the second highest value.
- **Bronze:** 50% of customers with lowest value.

A detailed methodological approach for the development of value-based segmentation will be presented in "A Guide for Value-Based Segmentation".

BEHAVIORAL SEGMENTATION

In behavioral segmentation the segments are identified with the application of appropriate clustering models on usage/behavioral data that usually reside in the organization's data warehouse or data marts. Thus behavioral segmentation can be implemented with a high degree of confidence and relatively low cost. Attributes that can be used for behavioral segmentation include product ownership and utilization, volume/type/frequency of transactions, payment and revenue history, and so on.

Typical behavioral segments that can be found in *banking* include:

- **Depositors:** Savings products – mostly deposit transactions using the network of branches.
- **Future investors:** Insurance and investment products – few payment and deposit transactions.
- **Consuming borrowers:** Consumer lending products (credit cards and consumer loans) – moderate to many transactions using all channels.
- **Frequent travelers:** All kinds of products – many transactions through different channels and many international transactions.
- **Shoppers:** Credit cards and other products – many transactions using mostly credit cards for purchases.
- **Needs borrowers:** Mortgage loans and consumer loans – mostly payment transactions using the network of branches.
- **Classic users:** Savings products and cards – moderate transactions mostly through branches and ATMs.
- **Transactioners:** Payroll savings products with low balances – many transactions mostly for making small withdrawals for everyday needs.
- **Inactive:** Unused savings accounts – no transactions.

Typical behavioral segments that can be found in *mobile telephony* include:

- **Roamers:** Heavy users of all available services – the key differentiating factor is that they use their cell/mobile phones to make calls from abroad.
- **Superstars:** Heavy users of all available services and all new cellular services (Internet, MMS, 3G, etc.).
- **Professional users:** Heavy voice users – increased voice usage and a very high incoming community (the incoming community is the number of distinct callers that have called the specific customer).

- **Classic users:** Average voice and SMS usage.
- **Youth – SMS users:** Heavy SMS users – they prefer using SMS to voice.
- **Oldies – basic users:** Voice usage only – very low incoming community.
- **Inactive:** No outgoing usage for a significant time period.

A detailed methodological approach for behavioral segmentation is presented in "A Guide for Behavioral Segmentation".

PROPENSITY-BASED SEGMENTATION

Propensity-based segmentation utilizes the results of classification models such as churn or cross- and up-selling models. This type of segmentation involves simple computations and the binning of customers in groups according to their propensity scores. For instance, customers can be divided into groups of low, medium, and high churn likelihood as a result of a churn model. Churn models estimate and assign churn propensity scores to all customers. These scores denote the likelihood of defection and enable marketers to rank customers according to their churn risk. When the churn prediction model has been developed, customers are sorted according to their churn scores. Appropriate threshold values are determined and customers are assigned to low-, medium-, or high-risk groups according to their estimated churn likelihood.

Analysts can combine multiple propensity models and scores to create compound segmentation schemes. This procedure typically requires the application of a clustering model. Once the propensity scores are estimated by the relevant models, they can be used as inputs fields in the clustering model. As an example let us consider the case of multiple propensity models that estimate the likely uptake of the following banking products:

- Savings products
- Investment products
- Credit cards
- Consumer loan
- Mortgage loan.

Cluster analysis can be applied to the respective propensity scores in order to reveal the segments of customers with similar future needs for specific product bundles. This approach may provide valuable support in the development of an optimized product bundling and cross-selling strategy.

Propensity scores and respective segmentations can also be combined with other standard segmentation schemes such as value-based segments. For instance,

Example segmentation

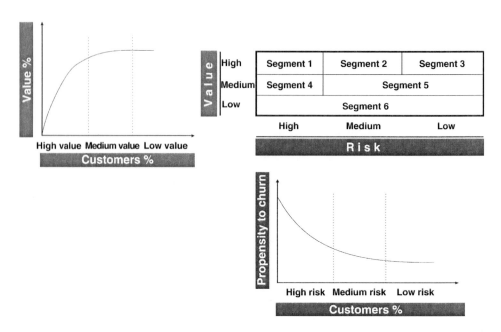

Figure 5.1 Combining value and churn propensity scores in value-at-risk segmentation.

when value segments are cross-examined with churn probability segments we have value-at-risk segmentation, a compound segmentation which can help in prioritizing the churn prevention campaign according to each customer's value and risk of defection. An example of this segmentation is shown in Figure 5.1. Nine compound segments are created after combining the three low-, medium-, and high-value segments with the three low-, medium-, and high-risk segments.

The first segment is the most critical one since it contains high-value customers at risk of defection. For this segment retention is the first priority.

LOYALTY SEGMENTATION

Loyalty segmentation is used to identify different groupings of customers according to their loyalty status and to separate loyals from migrators/switchers. The segments are created by the application of simple business rules and/or cluster models on survey or database information. By examining the loyalty segments an organization can gain insight into its strengths and weaknesses. A more elaborate loyalty segmentation is depicted in Figure 5.2 where loyals and migrators are further segmented according to the main reason for loyalty or migration.

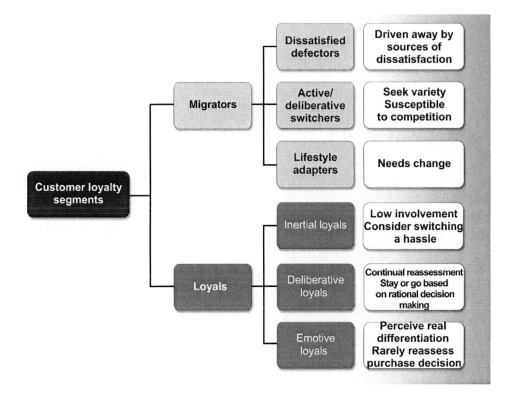

Figure 5.2   Loyalty segments.

Loyalty segments can be associated with specific usage behaviors and customer database attributes. To achieve this, an organization can start with a market survey to reveal the loyalty segments and then build a classification model with the loyalty segments' field as the target. That way, it will be able to identify the behaviors associated with each loyalty segment and use the relevant classification rules to extrapolate the loyalty segmentation results to the entire customer base.

For example, the behaviors identified as related to the customer loyalty segments of a mobile telephony company can be:

- **Dissatisfied defectors:** Complaints made to call center.
- **Active switchers:** Always with the optimum program/product, frequent changes of programs/products in search of optimization, new customers.
- **Lifestyle adapters:** Usage pattern changes that could indicate life-stage changes.
- **Inertial loyals:** Always with the same program/product, no changes at all even if the program/product that they use does not fit their usage profile.

- **Deliberative loyals:** Always with the optimum program/product, normal changes of programs/products in search of optimization, old customers.
- **Emotive loyals:** Loyal but not included with either the inertials or deliberatives.

SOCIO-DEMOGRAPHIC AND LIFE-STAGE SEGMENTATION

In demographic segmentation customers are grouped according to their demographics. It is a widely used segmentation since demographics are considered to have a strong influence on the customers' needs, preferences, and consuming/usage behaviors. We should point out, though, that in many cases people in the same demographic group may have different wants and needs as customers. The segments are typically created with simple business rules on the following customer attributes:

- Gender
- Race
- Social status
- Education
- Age
- Marital status
- Home ownership
- Family income
- Number of children/family size
- Age of children
- Occupation.

The family lifecycle segmentation is mainly determined by customers' age, marital status, and number/age of children. Typical family lifecycle segments include:

- Young, singles
- Young families/no children
- Young families/children under 5
- Growing families/children above 5
- Families with children
- Older retired persons.

Quite often, the customers' demographic data available in the organization's databases are limited, and/or outdated or of questionable validity. Organizations can augment their demographic data repository by purchasing data from an external supplier, provided of course that this is permitted by the relevant legislation. Alternatively, the data can be collected by a market survey.

In the data mining framework, demographic segmentation is mainly used to enhance insight into the revealed behavioral, value-, and propensity-based segments.

The life-stages are defined by special events that determine a person's priorities and main concerns, such as the birth of their first child, a significant increase in income, and so on. Life-stages present opportunities for promoting products and services that address the particular needs of customers.

An organization should try to identify the important life-stage events and link them to consuming behaviors. For example, in banking, the birth of a child is strongly related to consumer and/or mortgage loan purchases. A significant increase in annual income almost always creates investment and insurance needs and opportunities.

Since quite often the fields required for life-stage segmentation are not available in the organization's customer databases, analysts usually try to link the age of customers to specific life-stage events.

An example from the banking industry is shown in Figure 5.3.

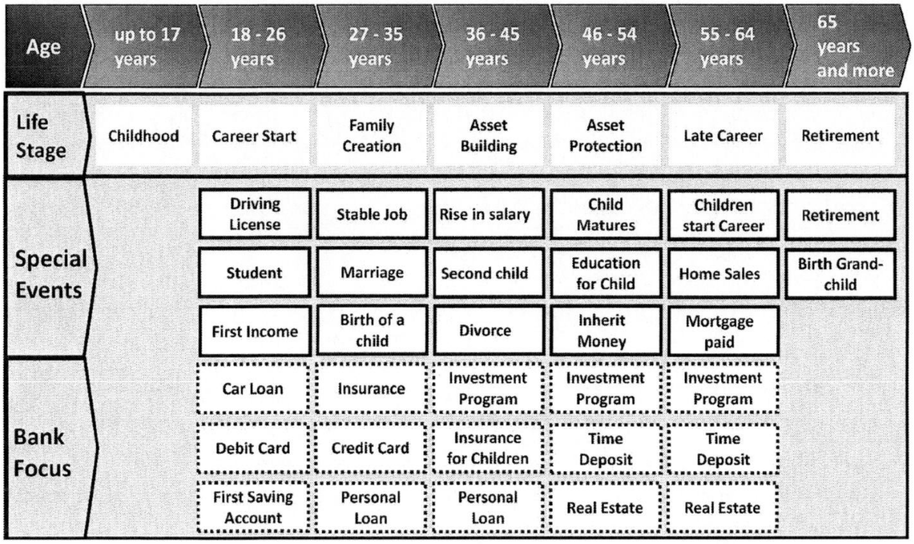

Age	up to 17 years	18 - 26 years	27 - 35 years	36 - 45 years	46 - 54 years	55 - 64 years	65 years and more
Life Stage	Childhood	Career Start	Family Creation	Asset Building	Asset Protection	Late Career	Retirement
Special Events		Driving License	Stable Job	Rise in salary	Child Matures	Children start Career	Retirement
		Student	Marriage	Second child	Education for Child	Home Sales	Birth Grand-child
	First Income	Birth of a child	Divorce	Inherit Money	Mortgage paid		
Bank Focus		Car Loan	Insurance	Investment Program	Investment Program	Investment Program	
		Debit Card	Credit Card	Insurance for Children	Time Deposit	Time Deposit	
		First Saving Account	Personal Loan	Personal Loan	Real Estate	Real Estate	

Figure 5.3   Life-stages in banking.

NEEDS/ATTITUDINAL-BASED SEGMENTATION

Needs/attitudinal-based segmentation is used to investigate customers' needs, wants, attitudes, perceptions, and preferences. Relevant information can only be collected through market surveys and the segments are typically identified by the application of a cluster model on gathered questionnaire responses.

In the data mining framework, needs/attitudinal segmentation is mainly used in combination with behavioral and value-based segments to enrich the profile of the revealed segments, provide insight into their "qualitative" characteristics, and hence support:

- New product design/development
- The communication of the product features/brand image important for each segment
- Tailored communication messages
- New promotions.

SEGMENTATION IN BUSINESS MARKETS

Consumers and business customers have inherent differences. A business market has fewer customers and larger transactions. A decision maker, who is typically not the actual user, makes decisions for a large number of users. Selling is a long and complex process and involves talks and negotiations with people who are not the end users. Furthermore, quite often there is little association between satisfaction level and customer loyalty.

In business markets almost every customer needs a customized product, quantity, or price. In fact, each customer can be considered as a distinct segment. The segmented marketing that successfully works in consumer markets is not effective in business markets where a one-to-one marketing approach is appropriate.

These fundamental differences impose a different segmentation approach for the business markets and the use of different segmentation criteria. For instance, segmenting customers by their behavioral and usage characteristics and communicating the product features important for each segment have limited effectiveness with business customers.

The segmentation criteria typically used for business customers include:

- Value (revenue or profit)
- Size (number of employees, number of subscriptions/subscribed employees, etc.)
- Industry (government, education, telecommunications, financial, etc.)
- Business life-stage (new business, mature business, etc.).

More specifically:

1. **Value based:** Customer value is one of the most important segmentation criteria for both consumer and business markets. In business markets, "customer" value is typically measured at multiple levels:

   (a) Per employee/subscription level for all subscribed employees within one corporate customer.
   (b) Per corporate (corporate account) level.
   (c) Average profit of all subscriptions/employees within one corporate customer.
   (d) Per industry.

   Consequently, value segmentation is also tackled at multiple levels. Usually, the differences in terms of value among corporate customers are quite large. The need to build and establish loyalty among the top-value customers is critical since, in general, they account for a substantial part of the organization's total profit.

2. **Size based:** Business customers are usually ranked and grouped according to their size, based on either their total number of employees or their total number of subscriptions (subscribed employees). Table 5.2 describes a relevant size-based segmentation scheme.

3. **Industry based:** Corporate customers can also be divided according to their industry with the use of a standard industrial classification system. A relevant high-level industrial categorization is listed in Table 5.3.

Table 5.2 Size-based segments in business markets.

Size category	Size-based segment	Definition
Large	• Multinational accounts • Government accounts • Corporate	• Multinational • Government • Number of employees greater than 200
Medium	• Business • SME (Small and Medium Enterprise)	• Number of employees between 50 and 200 • Number of employees between 10 and 50
Small	• SOHO (Small Office, Home Office) • Self-employed	• Number of employees up to 10 • Independent professionals – freelancers

Table 5.3   Industry-based segments in business markets.

Technology offensive industries:
IT/telecom
Bank/finance/insurance
Business services/real estate
Media
Shipping/oil/offshore
Interest organizations/others
Technology defensive industries:
Construction
Transportation
Primary industries
Manufacturing and power supply
Trade/shop/hotel/restaurant
Public sector:
Research and development
Health and social services
Public administration/police/defense

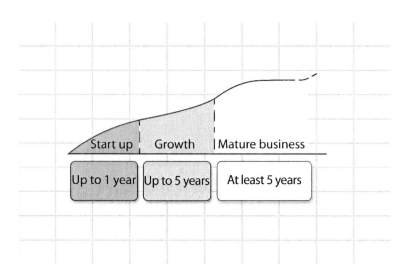

Figure 5.4   Typical life-stages of a business.

4. **Business life-stage:** Corporate customers can also be segmented according to their business life-stage, and the design of the marketing strategy should be differentiated accordingly. Typical business life-stages include the startup, growth, and maturity phases (Figure 5.4).

A GUIDE FOR BEHAVIORAL SEGMENTATION

A segmentation project starts with the definition of the business objectives and ends with the delivery of differentiated marketing strategies for the segments. In this section we will focus on behavioral segmentation and present a detailed methodological approach for the effective implementation of such a project.

This section also includes guidelines for the successful deployment of the segmentation results and presents ways for using the revealed segments for effective marketing.

BEHAVIORAL SEGMENTATION METHODOLOGY

The proposed methodology for behavioral segmentation includes the following main steps which are presented in detail in the following sections:

1. **Business understanding and design of the segmentation process.**
2. **Data understanding, preparation, and enrichment.**
3. **Identification of the segments with cluster modeling.**
4. **Evaluation and profiling of the revealed segments.**
5. **Deployment of the segmentation solution, design, and delivery of the differentiated strategies.**

The sequence of stages is not rigid. Lessons learned in each step may lead analysts to review previous steps.

Business Understanding and Design of the Segmentation Process

This phase starts with understanding the project requirements from a business perspective. It involves knowledge-sharing meetings and close collaboration between the data miners and marketers involved in the project to assess the situation, clearly define the specific business goal, and design the whole data mining procedure. In this phase, some crucial questions must be answered and decisions on some very important methodological issues should be taken. Tasks in this phase include:

1. **Definition of the business objective:** The business objective will determine all the next steps in the behavioral segmentation procedure. Therefore it should be clearly defined and translated to a data mining goal. This translation

of a marketing objective to a data mining goal is not always simple and straightforward. However, it is one of the most critical issues, since a possible misinterpretation can result in failure of the entire data mining project.

It is very important for the analysts involved to explain from the outset to everyone involved in the data mining project the anticipated final deliverables and to make sure that the relevant outcomes cover the initially set business requirements.

2. **Selection of the appropriate segmentation criteria:** One of the key questions to be answered before starting the behavioral segmentation is what attributes are to be used for customer grouping. The selection of the appropriate segmentation criteria depends on the specific business issue that the segmentation model is about to address. The business needs imply, if not impose, the appropriate input fields. Usually people with domain knowledge and experience can provide suggestions on the key attributes related to the business goal of the analysis. All relevant customer attributes should be identified, selected, and included in the segmentation process. Information not directly related to the behavioral aspects of interest should be omitted.

For instance, if a mobile telephony operator wants to group its customers according to their use of services, all relevant fields, such as the number and volume/minutes of calls by call type, should be included in the analysis. On the contrary, customer information related to other aspects of customer behavior, such as payment or revenue information, should be excluded from the segmentation.

3. **Determination of the segmentation population:** This task involves the selection of the customer population to be segmented. An organization may decide to focus on a specific customer group instead of the entire customer base. In order to achieve more refined solutions, groups that have apparent differences, such as business or VIP customers and typical consumer customers, are usually handled by separate segmentations.

Similarly, customers belonging to obvious segments, such as inactive customers, should be set apart and filtered out from the segmentation procedure in advance. Otherwise, the large differences between active and inactive customers may dominate the solution and inhibit identification of the existing differences between active customers.

If the size of the selected population is large, a representative sample could be selected and used for model training. In that case, though, a deployment procedure should be designed, for instance through the development of a relevant classification model, which will enable the scoring of the entire customer base.

4. **Determination of the segmentation level:** The segmentation level defines what groupings are about to be revealed, for instance groups of customers,

groups of telephone lines (MSISDNs in mobile telephony), and so on. The selection of the appropriate segmentation level depends on the subsequent marketing activities that the segments are about to support. It also determines the aggregation level of the modeling dataset that is going to be constructed.

Data Understanding, Preparation, and Enrichment

The investigation and assessment of the available data sources is followed by data acquisition, integration, and processing for the needs of segmentation modeling. The data understanding and preparation phase is probably the most time-consuming phase of the project and includes tasks such as:

1. **Data source investigation:** The available data sources should be evaluated in terms of accessibility and validity. This phase also includes initial data collection and exploration in order to understand the available data.
2. **Defining the data to be used:** The next step in the procedure involves the definition of the data to be used for the needs of the analysis.

    The selected data should cover all the behavioral dimensions that will be used for the segmentation as well as all the additional customer information that will be used to gain deeper insight into the segments.
3. **Data integration and aggregation:** The initial raw data should be consolidated to create the final modeling dataset that will be used for identification of the segments. This task typically includes the collection, filtering, merging, and aggregation of the raw data. But first the structure of the modeling dataset should be defined, including its contents, time frame of used data, and aggregation level.

    For behavioral segmentation applications, a recent "view" of the customers' behavior should be constructed and used. This "view" should summarize the behavior of each customer by using at least six months of recent data (Figure 5.5).

    The aggregation level of the modeling dataset should correspond to the required segmentation level. If the goal, for instance, is to segment bank customers, then the final dataset should be at a customer level. If the goal is to segment telephone lines (MSISDNs), the final dataset should be at a line level. To put it in a simpler way, clustering techniques reveal natural groupings of records. So no matter where we start from, the goal is the construction of a final, one-dimensional, flat table, which summarizes behaviors at the selected analysis level.

    This phase concludes with the retrieval and consolidation of data from multiple data sources (ideally from the organization's mining data mart and/or MCIF) and the construction of the modeling dataset.

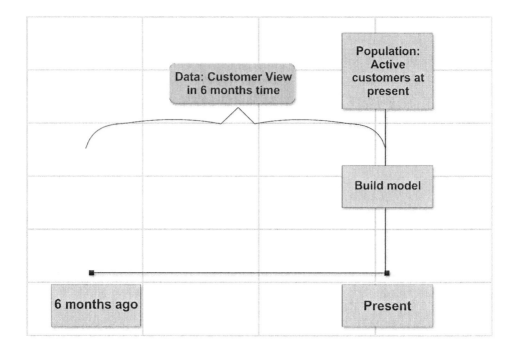

Figure 5.5 Indicative data setup for behavioral segmentations.

4. **Data validation and cleaning:** A critical issue for the success of any data mining project is the validity of the used data. The data exploration and validation process includes the use of simple descriptive statistics and charts for the identification of inconsistencies, errors, missing values, and outlier (abnormal) cases. Outliers are cases that do not conform to the patterns of "normal" data. Various statistical techniques can be used in order to fill in (impute) missing or outlier values. Outlier cases in particular require extra care. Clustering algorithms are very sensitive to outliers since they tend to dominate and distort the final solution. For general purpose behavioral segmentations, the outlier cases can also be filtered out so that the effect of "noisy" records in the formation of the clusters is minimized.

   Problematic values, particularly demographic information, can also be imputed or replaced by using external data, provided of course the external data are legal, reliable, and can be linked to the internal data sources (e.g., through the VAT number, post code, phone number, etc.).

5. **Data transformations and enrichment:** This phase deals with the enrichment of the modeling dataset with derived fields such as ratios, percentages, averages, and so on. The derived fields are typically created by the application

of simple functions on the original fields. Their purpose is to better summarize customer behavior and convey the differentiating characteristics of each customer. This is a critical step that depends greatly on the expertise, experience, and "imagination" of the project team since the development of an informative list of inputs can lead to richer and more refined segmentations.

The modeling data may also require transformations, specifically standardization, so that the values and the variations of the different fields are comparable. Clustering techniques are sensitive to possible differences in the measurement scale of the fields. If we do not deal with these differences, the segmentation solution will be dominated by the fields measured in larger values. Fortunately, many clustering algorithms offer integrated standardization methods to adjust for differences in measurement scales. Similarly, the application of a data reduction technique like principal components analysis (PCA) or factor analysis also provides a solution, since the generated components or factors have standardized values.

6. **Data reduction using PCA or factor analysis:** The data preparation stage is typically concluded by the application of an unsupervised data reduction technique such as PCA or factor analysis. These techniques reduce the data dimensionality by effectively replacing a typically large number of original inputs with a relatively small number of compound scores, called factors or principal components. They identify the underlying data dimensions by which the customers will be segmented. The derived scores are then used as inputs in the clustering model that follows. The advantages of using a data reduction technique as a data preprocessing step include:

(a) Simplicity and conceptual clarity. The derived scores are relatively few, interpreted, and labeled. They can be used for cluster profiling to provide the first insight into the segments.
(b) Standardization of the clustering inputs, a feature that is important in yielding an unbiased solution.
(c) Equal contributions from the data dimensions to the formation of the segments.

Factor Analysis Technical Tips

PCA is the recommended technique when the primary goal is data reduction.

In order to simplify the explanation of the derived components, the application of a rotation, typically Varimax, is recommended.

The most widely used criterion for deciding the number of components to extract is "Eigenvalues over 1." However, the final decision should also take into account the percentage of the total variance explained by the extracted components. This percentage should, in no case, be lower than 60–65%.

Most importantly, the final components retained should be interpretable and useful from a business perspective. If an extra component makes sense and provides conceptual clarity then it should be considered for retention, as opposed to one that makes only "statistical" sense and adds nothing in terms of business value.

Identification of the Segments with Cluster Modeling

Customers are divided into distinct segments by using cluster analysis. The clustering fields, typically the component scores, are fed as inputs into a cluster model which assesses the similarities between the records/customers and suggests a way of grouping them. Data miners should try a test approach and explore different combinations of inputs, different models, and model settings before selecting the final segmentation scheme.

Different clustering models will most likely produce different segments and this should not come as a surprise. Expecting a unique and definitive solution is a sure recipe for disappointment. Usually the results of different algorithms are not identical but similar. They seem to converge to some common segments. Analysts should evaluate the agreement level of the different models and examine which aspects disagree. In general, a high agreement level between many different cluster models is a good sign for the existence of discernible groupings.

The modeling results should be evaluated before selecting the segmentation scheme to be deployed. This takes us to the next stage of the behavioral segmentation procedure.

Evaluation and Profiling of the Revealed Segments

In this phase the modeling results are evaluated and the segmentation scheme that best addresses the needs of the organization is selected for deployment. Data miners should not blindly trust the solution suggested by one algorithm. They should explore different solutions and always seek guidance from the marketers for selecting the most effective segmentation. After all, they are the ones who will use the results for segmented marketing and their opinion on the future benefits of each solution is critical. The selected solution should provide distinct and meaningful clusters that can indicate profitable opportunities. Tasks in this phase include:

1. **"Technical" evaluation of the clustering solution:** The internal cohesion and separation of the clusters should be assessed with the use of descriptive statistics and specialized technical measures (standard deviations, interclass and intraclass distances, silhouette coefficient, etc.) such as the ones presented in the previous chapter. Additionally, data miners should also examine the distribution of customers in the revealed clusters as well as consistency of the results in different datasets. All these tests assess the segmentation solution in terms of "technical" adequacy. Additionally, the segments should also be assessed from a business perspective in terms of actionability and potential benefits. To facilitate this evaluation, a thorough profiling of the segments' characteristics is needed.

2. **Profiling of the revealed segments:** A profiling phase is typically needed in order to fully interpret the revealed segments and gain insight into their structure and defining characteristics. Profiling supports the business evaluation of the segments as well as the subsequent development of effective marketing strategies tailored for each segment.

   Segments should be profiled by using all available fields as well as external information. The description of the extracted segments typically starts with the examination of the centroids table. The centroid of each cluster is a vector defined by the means of its member cases on the clustering fields. It represents the segment's central point, the most representative case of the segment. The profiling phase also includes the use of simple reporting and visualization techniques for investigating and comparing the structures of the segments. All fields of interest, even those not used in the formation of the segments, should be cross-examined with them to gain deeper insight into their meaning. However, analysts should always look cautiously at the demographics of the derived segments. In many cases the demographic information may have not been updated since the customer's first registration. Analysts should also bear in mind that quite often the person using the service (credit card, mobile phone, etc.) may not be the same one registered as a customer.

3. **Cluster profiling with supervised (classification) models:** Classification models can augment reporting and visualization tools in the profiling of the segments. The model should be built with the segment assignment field as the target and the profiling fields of interest as inputs. Decision trees in particular, due to the intuitive format of their results, are typically used to outline the segment profiles.

4. **Using marketing research information to evaluate and enrich the behavioral segments:** Marketing research surveys are typically used to investigate the needs, preferences, opinions, lifestyles, perceptions, and attitudes of the customers. They are also commonly used in order to collect valid and updated demographic information. It is strongly recommended that the data mining-driven behavioral segments are combined with the market research-driven

demographic and/or needs/attitudinal segments. While each approach helps with understanding certain aspects of the customers, combining the approaches provides deeper insight.

For instance, provided a behavioral data mining segmentation has been implemented, random samples can be extracted from each segment and, through surveys and/or qualitative research and focus group sessions, valuable insight can be gained concerning each segment's needs and preferences. Alternatively, the data mining and the market research approaches can be implemented independently and then cross-examined, not only as a means for evaluating the solutions, but also in order to construct a combined and integrated segmentation scheme which would provide a complete view of the customers.

In conclusion, combining data mining and market research techniques for customer segmentation can enable refined subsequent marketing strategies, based on a thorough understanding of customer behavior and needs, as shown in Figure 5.6.

Consider customers belonging to the same behavioral segment but having diverse needs and perceptions. This information can lead to tailored marketing strategies within each behavioral segment.

5. **Labeling the segments based on their identified profiles:** The profiling and interpretation process ends with the labeling of the identified segments with names that appropriately designate their unique characteristics. Each segment is assigned an informative and revealing name, for instance "Business Travelers," instead of "Segment 1." The naming of the segments should take into account all the profiling findings. These names will be used for communicating the segments to all business users and for loading them onto the organization's operational systems.

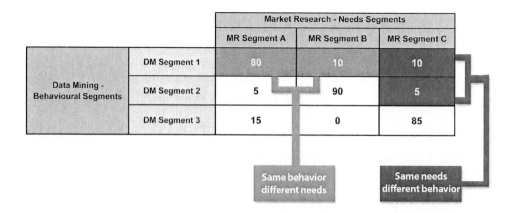

Figure 5.6   Combining data mining and market research-driven segmentations.

Deployment of the Segmentation Solution, Design, and Delivery of Differentiated Strategies

The segmentation project concludes with the deployment of the segmentation solution and its use in the development of differentiated marketing strategies and segmented marketing:

1. **Building the customer scoring model for updating the segments:** The deployment procedure should enable customer scoring and updating of the segments. It should be automated and scheduled to run frequently to enable the monitoring of the customer base over time and the tracking of segment migrations. Moreover, because nowadays markets change very rapidly, it is evident that a segmentation scheme can become outdated within a short time. Refreshment of such schemes should be made quite often. This is why the developed procedure should also take into account the need for possible future revisions.
2. **Building a decision tree for scoring – fine tuning the segments:** A decision tree model can be used as a scoring model for assigning customers to the revealed clusters. The derived classification rules are understandable and provide transparency compared to scoring with the cluster model. More importantly, business users can easily examine them and possibly modify them to fine-tune the segmentation based on their business expertise. This combined approach can create stable segments unaffected by seasonal or temporary market conditions.
3. **Distribution of the segmentation information:** Finally, the deployment procedure should also enable distribution of the segmentation information throughout the enterprise and its "operationalization." Therefore it should cover uploading of the segmentation information to the organization's databases, as well as operational systems, in order to enable customized strategies to be applied across all customers' touch points.

TIPS AND TRICKS

The following tips should be taken into account when planning and carrying out a segmentation project:

- Take into account the core industry segments before proceeding with the segmentation project and then decide which core segments need further analysis and sub-segmentation.

- Clean your data from obvious segments (e.g., inactive customers) before proceeding with the segmentation analysis.
- Always bear in mind that eventually the resulting model will be deployed. In other words, it will be used for scoring customers and for supporting specific marketing actions. Thus, when it comes to selecting the population to be used for model training, do not forget that this is the same population that will be scored and included in a marketing activity at the end. So sometimes it is better to start with the end in mind and consider who we want to score, segment, or classify at the end: the entire customer base, consumer customers, only high-value customers, and so on. This deployment-based approach can help us to resolve ambiguities about selection of the modeling dataset population.
- Select only variables relevant to the specific business objective and the particular behavioral aspects you want to investigate. Avoid mixing all available inputs in an attempt to build a "magic" segmentation that will cover all aspects of a customer's relationship with the organization (e.g., phone usage and payment behavior).
- Avoid using demographic variables in a behavioral segmentation project. Mixing behavioral and demographical information may result in unclear and ambiguous behavioral segments since two customers with identical demographic profiles may have completely different behaviors.
- Consider the case of a father who has activated a mobile phone line for his teenage son. In a behavioral segmentation solution, based only on behavioral data, this line would most likely be assigned to the "Young – SMS users" segment, along with other teenagers and young technophile users. Therefore we might expect some ambiguities when trying to examine the demographic profile of the segments. In fact, this hypothetical example also outlines why the use of demographic inputs should be avoided when the main objective is behavioral separation.
- 'Smooth' your data. Prefer to use monthly averages, percentages, ratios, and other summarizing KPIs that are based on more than one month of data.
- A general recommendation on the time frame of the behavioral data to be used is to avoid using less than 6 months and more than 12 months of data in order to avoid founding the segments on unstable/volatile or outdated behaviors.
- Try different combinations of input fields and explore different models and model settings. Build numerous solutions and pick the one that best addresses the business goal.
- Labeling the segments needs extra care. Bear in mind that this label will characterize the segments, so a hasty naming will misguide all recipients/users of this information. A plain label will unavoidably be a kind of derogation, as it is impossible to incorporate all the differentiating characteristics of a segment. Yet, a carefully chosen name may simply and successfully communicate the unique characteristics of the segments to all subsequent users.

- Always prefer supervised to unsupervised models when your business problem concerns the classification of customers into categories known in advance. Predicting events (e.g., purchase of a product, churn, defaulting) is a task that can be addressed much more efficiently by classification models. A decision tree that estimates the event's propensities will produce better results than any unsupervised model.
- Customers that do not fit well to their segment should be set apart and assigned to an "unclassified" segment in order to improve the homogeneity and quality of the segments. For instance, a customer with a very low amount in stocks and a stocks-only product portfolio, even if classified as an "Investor" by a clustering algorithm, should be identified and assigned either to the "Unclassified" or to the "Passive/Inactive" group.

SEGMENTATION MANAGEMENT STRATEGY

The following roadmap (Figure 5.7) outlines the main steps that a segmentation-oriented enterprise should follow to develop an effective marketing strategy based on behavioral segments:

Step 1: Identify the customer segments in the database

Initially the segmentation results should be deployed in the customer database. Customers should be scored and assigned into segments.

In this phase the involved team should:

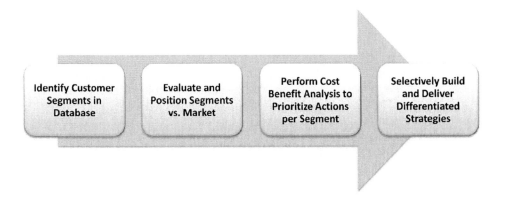

Figure 5.7 Roadmap of the segmentation management strategy.

- Fine tune the scoring procedure, assigning customers to segments.
- Create size and profiling reports for each segment.

Step 2: Evaluate and position the segments

This step involves the evaluation and positioning of the segments in the market. The possible opportunities and threats should be investigated in order to select the key segments to be targeted with marketing activities tailored to their profiles and needs.

In this phase the involved team should:

- Conduct a market survey to enrich segments with qualitative information such as:

  - Needs and wants
  - Lifestyle and social status.

- Analyze segments and:

  - Identify competitive advantages.
  - Identify opportunities and threats.
  - Set marketing and financial goals.
  - Select the appropriate marketing mix strategies: product, promotion, pricing, distribution channels.

- Select and define the appropriate KPIs for monitoring the performance of the segments

Step 3: Perform cost-benefit analysis to prioritize actions per segment

Effective segmentation management requires insight into the factors that drive customer behaviors. Behavioral changes, such as customer development/ additional purchases and attrition, should be analyzed through data mining and market research. The important drivers for each segment should be recognized and the appropriate marketing actions/interventions that can have a positive impact on the customer relationship with the organization should be identified. Finally, a cost–benefit analysis can enable the prioritization of the marketing actions to be implemented.

The analysis per segment should include:

- Identification of the main factors of profitability
- Identification of the main cost factors
- Assessment of the importance of behavioral drivers.

Identifying and Quantifying the Significant Behavioral Drivers

In CRM a deep understanding of customer attrition and development is critical. This understanding should involve the identification of factors that drive behavioral changes and have an effect on the decision to buy or churn.

When a segmentation scheme is in place, evaluation of the drivers' importance and impact should be carried out separately for each segment since the effect of the drivers can vary from segment to segment.

The basic drivers that should be examined in relation to the decision to buy or to churn include:

- Properties of products/services
- Price of products/services
- Properties of contract
- Products/services use
- Change in needs/life-stage
- Technology needs
- Processes' efficiency
- Loyalty programs
- Branch/POS network
- Proactive contact with relevant offers
- Service satisfaction
- Brand image.

Step 4: Build and deliver differentiated strategies

In order to make the most of segmentation, specialization is needed. Thus, it is suggested that each segment is addressed by a different segment management team. Each team should build and deliver specialized marketing strategies to improve customer handling, develop the relationship with customers, and increase the profitability of each segment. The responsibilities of the team should also include the tracking of each segment with appropriate KPIs and the monitoring of competitors' activities for each segment.

Segment management teams should take into account the different demands of each segment and work separately as well as together to design strategic marketing plans focusing on the following:

- **Channel management:** Offer the appropriate channel mix in order to cover each segment's particular demands.

- **Marketing services management:** Optimize the effectiveness of marketing campaigns.
- **Product management:** Manage existing product offerings and new product development.
- **Brand management:** Understand customer perception of the brand, and identify and communicate the appropriate brand image and key product benefits.

A GUIDE FOR VALUE-BASED SEGMENTATION

In contrast to behavioral segmentation, which is multi-attribute since it typically involves the examination of multiple segmentation dimensions, value-based segmentation is one dimensional as customers are grouped on the basis of a single measure, the customer value. The most difficult task in such a project is the computation of a valid value measure for dividing customers, rather than the segmentation itself.

VALUE-BASED SEGMENTATION METHODOLOGY

As with any other data mining project, the implementation of value segmentation should be guided by a specific methodology, a core process model, in order to achieve an effective outcome. Inevitably the core process model for value-based segmentation has many similarities to the one described in the previous sections for the needs of behavioral segmentation. In this section we will outline the main methodological steps of the value segmentation, emphasizing the special tasks related to such projects.

Business Understanding and Design of the Segmentation Process

This phase includes the setting of the business goal, an assessment of the situation, and the design of the process. Those analysts and marketers engaged in the project should co-operate with finance department officers to jointly determine the appropriate value measure that will be used to divide the customer base. This co-operation typically involves the identification and explanation/understanding of the cost and revenue components and a preliminary assessment of their availability. It should finally conclude with agreement on a formula that will be used for calculating the value measure.

More sophisticated measures are more effective, but harder to construct and deploy, and have a greater risk of failure. Potential value measures include:

1. **ARPU (Average Revenue Per Unit):** A revenue-only measure which does not take into account the different product margins, the costs to manage, and the expected customer lifetime (length of relationship with the organization).
2. **MARPU (Marginal Average Revenue Per Unit):** Revenue minus the cost of providing the products and services used.
3. **Customer profitability:** Cost to manage is also taken into account.
4. **Potential value:** The expected profit a customer will generate through new purchases.
5. **LTV (Lifetime Value):** Estimation of customer's future value which also takes into account the customer's expected lifetime and potential value from new purchases (LTV = customer present value + potential value).

As an example, let us consider Table 5.4, which presents a list of factors, costs, and revenues that should contribute to the development of a MARPU index in mobile telephony.

Table 5.4 Profitability factors in mobile telephony.

Factor	Description	MARPU contribution
Monthly access	Monthly access fees	+
Other one-off fees	Fees that a subscriber pays once (e.g., change of SIM card)	+
GSM revenue	Revenue from GSM usage	+
GPRS revenue	Revenue from GPRS usage	+
Other VAS revenue	Revenue from other VAS	+
Incoming revenue	Revenue from incoming calls	+
Volume discount	Discount given to the subscriber	−
Customer discount	Discount given to the subscriber	−
Interconnection costs	Cost from interconnection costs between operators	−
Termination costs	Cost from termination costs to same network	−
Roaming costs	Cost from roaming calls	−
Services costs	Cost from services offered to subscribers	−
Other dealer costs	Money paid to other dealers	−

A similar list of factors that should be taken into account in the development of a MARPU index in banking is presented in Table 5.5.

Table 5.5 Profitability factors in banking.

Factor	Description	MARPU contribution
Margin	The difference between the interest gained and the interest expense	+
Funding credit	The amount earned by banks on the deposits employed	+
Cost of funds	Amount of interest expense paid for the funds used by banks	−
Loan loss provision	The amount reserved by banks to cover expected loan losses	−
Capital charge	The capital allocated multiplied by rate of return	−
Overhead	The fixed and variable expenses associated with the product (e.g., transactional)	−

The tasks of this phase also include selection of the segmentation population and level as well as selection of the value cut-points that will be used to divide customers, for example, top 1%, high 19%, medium 40%, low 40%.

Data Understanding and Preparation – Calculation of the Value Measure

Data understanding includes a thorough investigation of the availability of the required data components and the detection of possible problems that may be encountered due to missing information. After that, all relevant customer information should be consolidated, evaluated in terms of quality, cleaned, and prepared for calculating a credible value measure. This phase ends with the development of an accurate and valid formula for quantifying the customers' value. The calculation should use input data that cover a sufficient period of time, at least six months, so that it accurately summarizes the value of each customer.

Grouping Customers According to Their Value

There is no modeling phase in value-based segmentation. Instead of modeling, this segmentation type requires simple computations for the calculation of a credible value measure and the use of this measure to divide customers accordingly.

Customers are sorted according to the calculated value measure and binned in groups of equal size, named quantiles, such as tiles of 1% (percentiles) or tiles of 10% (deciles). For instance, the partitioning in deciles would result in 10 groups of

10% each. The customers in tile 1, assuming a ranking of customers in descending order, would contain the top 10% of customers with the highest value index, whereas tile 10 would contain the 10% of customers with the lowest value.

The derived tiles may be combined to form refined segments of the form top n%, high n%, medium high n%, medium low n%, low n%, and so on. The number and size of the derived segments define the refinement level of the segmentation solution and depend on the specific needs of the organization. Customers with zero or even negative value contribution should be identified in advance, excluded from the binning procedure, and assigned to distinct groups.

Profiling and Evaluation of the Value Segments

The derived value measure should be cross-examined with other relevant customer attributes, including all the available cost and revenue figures, as a means of evaluating the validity of the calculation and the value index.

Analysts should examine the identified segments in terms of their contribution to the total profitability/revenue of the organization in order to assess their relative importance. Therefore the size, total number of customers, and total revenue/profit that each segment accounts for should be calculated and presented in tables and/or charts such as the ones shown below for a hypothetical segmentation scheme.

The contents of Table 5.6 are graphically illustrated in the profitability curve of Figure 5.8 which depicts the cumulative size and revenue of the derived segments of our hypothetical example. Quite often a substantial part of the organization's profit/revenue stems from a disproportionately small part of the customer base. The steep incline and the high "slope" of the curve on the left of the graph designate the large concentration of value on a relatively small percentage of customers.

Finally, the structure of the value segments should be investigated by examining all information of interest, including behavioral and demographic attributes. This profiling aims at identifying the defining characteristics of the value segments. It can also include a cross-examination of the value segments with other available segmentation schemes, such as behavioral, demographic, or needs/attitudinal segments. The organization should initially focus on an examination of high-value customers since their profile can provide insight that can be used to identify existing or prospective customers with high-value potential.

Deployment of the Segmentation Solution

The last step includes the design and development of a deployment procedure for updating the value segments. The procedure should be automated to allow customers' value to be tracked over time. It should also load the generated results into the organization's databases as well as its operational systems for subsequent use. The marketing usage of the value segments is presented next.

Table 5.6   A summary of the size and contribution of each value segment.

Value segment (tile)	% customers	Cumulative % customers	% revenue	Cumulative % of sum revenue
1	10	10	35	35
2	10	20	20	55
3	10	30	14	69
4	10	40	9	78
5	10	50	7	85
6	10	60	5	90
7	10	70	4	94
8	10	80	3	97
9	10	90	2	99
10	10	100	1	100

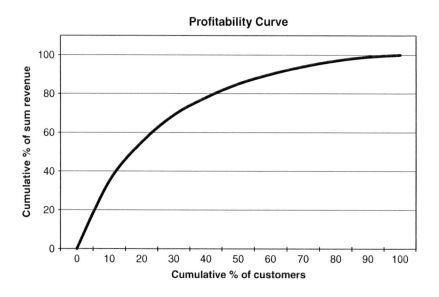

Figure 5.8   The profitability curve of the value segments.

DESIGNING DIFFERENTIATED STRATEGIES FOR THE VALUE SEGMENTS

The value-based segments should characterize each customer throughout his or her relationship with the organization and appropriate marketing objectives should be set for each segment.

Retention should be set as a top priority for high-value customers. In many cases a relatively small percentage of high-value customers accounts for a disproportionately large percentage of the overall company profit. Therefore maximum effort and allocation of resources should be made for preventing the defection of those customers. Additionally, the organization should try to further grow its relationship with them. Development through targeted cross- and up-selling campaigns should be set as a high priority for medium- and low-value customers with growth potential. At the same time, though, driving down the cost to serve should also be considered for those customers making trivial contributions to the organization's profitability. The main priorities by core value segment are illustrated in Figure 5.9.

Table 5.7 provides examples of how the value segments can be used to support the design of different strategies, the prioritization of the marketing interventions, and the delivery of differentiated service level.

Table 5.7 Differentiated treatment by value segments.

	Handling by customer service	Retention activities	Use of communication channel	Launch of a strategic product
High-value customers	Requests/ inquiries/ complaints/ interactions handled by the most experienced representatives, with top priority and with minimum "wait" time	Maximum effort and allocation of resources for retention and preventing churn. Close monitoring of their churn risk with the development of respective churn models. Offer of enticing incentives when at risk of defection	Identify and use the preferred communication channel	Communicated to all customers through their preferred channel
Medium-value customers	Their handling is second in priority compared to high-value customers	Moderate effort for retention and preventing churn	Use the preferred channel but also take the cost factor into account	Communicated to all customers mainly through their preferred channel

(continued overleaf)

Table 5.7   (*continued*)

	Handling by customer service	Retention activities	Use of communication channel	Launch of a strategic product
Core value customers	Last in the "queue" for customer service	Limited allocation of resources for preventing churn	Use the lowest cost channel	Communicated to all customers through the lowest-cost channel

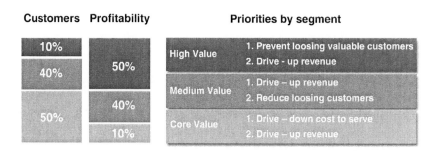

Figure 5.9   Main priorities by value segments.

Monitoring the value segments is vital. It should be an ongoing process integrated into the organization's procedures. Substantial upward or downward movements across the value "pyramid" (Figure 5.10) merit special investigation.

Segment migrations can be monitored through simple reporting techniques which compare the segment assignment of the customers over two distinct time periods (Figure 5.11), for instance at present and a few months back. This before–after type of analysis, also called cohort analysis, enables the organization to assess the evolution of its customer base.

Customers with substantial upward movement seem to strengthen their relationship with the organization. Finding their "clones" through predictive modeling and identifying customers with similar profiles can reveal other customers with growth potential.

At the other end, a decline in the pyramid suggests a relationship that is fading. This decline can be considered as a signal of a "leaving" customer. In many cases, for instance when there is not a specific event which signifies the termination of the relationship with the customer, a churn event cannot be explicitly defined through recorded data. In those cases, or as a way for being proactive, analysts can use a substantial value decline to define attrition and develop a respective churn model.

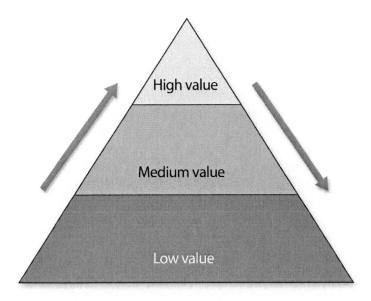

Figure 5.10   Migrations across value segments.

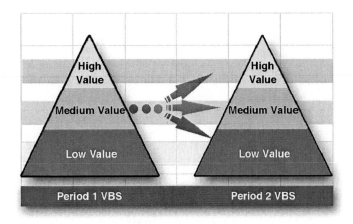

Figure 5.11   Value-based segment (VBS) migrations as a means of investigating the evolution of the customer base.

SUMMARY

Customer segmentation is the process of identifying groups that have common characteristics. The main objective of customer segmentation is to understand the customer base and gain customer insight that will enable the design and development of differentiated marketing strategies.

Clustering is a way to identify segments not known in advance and split customers into groups that are not previously defined. The identification of the segments should be followed by profiling the revealed customer groupings. Profiling is necessary for understanding and labeling the segments based on the common characteristics of the members.

The criteria used to divide customers (behavioral, demographic, value or loyalty information, needs/attitudinal data) define the segmentation type. In this chapter we briefly introduced some of the most widely used segmentation types before concentrating on behavioral and value-based segmentations which constitute the main topic of the book. For these types we proposed a detailed methodological framework and suggested ways for the use of their results in marketing. The following chapters feature a set of detailed case studies which show the use of the proposed methodology in real-world behavioral and value-based segmentation applications for various major industries.

CHAPTER SIX

Segmentation Applications in Banking

SEGMENTATION FOR CREDIT CARD HOLDERS

Credit card customers use their cards in diverse ways. They buy different products and services in order to cover their specific needs and according to their particular way of life, personality, and life-stage. Customers also vary in terms of their frequency and intensity of purchase. Frequent shoppers use their cards to cover their daily needs while occasional shoppers use them sporadically and only when they are about to charge the card with a large amount for a significant purchase. Additionally, one-off shoppers tend to make single purchases in contrast to installment buyers.

Since nowadays Internet shopping supplements the traditional trips to retail stores, credit card usage is also differentiated according to the preferred purchase channel.

Cash advances are another differentiating factor for credit card holders. A cash advance is the use of a credit card in ATMs to withdraw cash against the credit limit. This option is utilized by a specific part of the customer base as a personal loan alternative.

Payment time is unavoidable for all customers. However, even then not all customers behave in the same way. Some of them tend to pay the full amount due, whereas others tend to pay a minimum amount of their balance. The latter, usually referred to as revolvers, tend to carry past balances, as opposed to transactors who usually pay their full balances every month.

Although things are beginning to get quite complicated already, there is also good news. Credit card transactions are typically recorded and stored in relevant datasets, providing a complete footprint of customer behavior.

In the next sections we will follow the efforts of a bank to segment consumer credit card holders according to their usage behavior. The bank's objective was to reveal the underlying customer groups with similar usage characteristics in order to design differentiated marketing strategies.

The first decision that the involved marketers had to make concerned the main segmentation criteria by which they would group the credit card customer base. Their primary goal was to use segmentation in order to develop new card types and reward programs tailored to the different customer characteristics. Therefore they decided to start by focusing on the customers' purchasing habits and the products/services that they spend their money on. Their plan was to mine purchasing preferences and assign customers to segments according to the type of products that they tend to buy with their cards.

DESIGNING THE BEHAVIORAL SEGMENTATION PROJECT

The goal set by the organization's marketing officers was to group customers with respect to their purchasing habits and, more specifically, in terms of the mix of products/services that they tend to buy. This marketing objective was translated into a data mining goal of behavioral segmentation by using a clustering algorithm to automatically detect the natural groupings in the customer base. The final outcome of this project would be the identification of the different customer typologies and the assignment of each customer to a corresponding behavioral segment. The data chosen for segmentation purposes included the relative spending per merchant. This information was available in the mining data mart of the organization where credit card purchase transactions were differentiated by merchant category, enabling identification of the purchasing preferences of each customer.

The implementation procedure for the project also included the application of PCA to identify discrete data dimensions before clustering. Moreover, it involved a phase of thorough exploration of the derived clustering solution in order to interpret the clusters and evaluate the derived solution.

The organization's marketers wanted not only to identify the underlying customer groupings, but also to disseminate this information throughout the organization, including the front-line systems, to ensure its use in everyday marketing activities. Additionally, they wanted to be able to monitor the revealed segments over time. Therefore the project also included the development of a deployment procedure for easy updating of the segments that would also allow tracking of segment evolution over time.

Additional methodological issues that the project team had to tackle before beginning the actual implementation included:

- The level of merchant grouping that should be used in the analysis.
- The segmentation population and whether it should cover all customers and credit cards or only a subset of particular interest.

The answers to these issues and the reasoning behind the selected approach are presented next.

BUILDING THE MINING DATASET

The required usage data were available in the organization's mining data mart and MCIF, which constituted the main information sources for the needs of the project. The recorded data covered all aspects of credit card use including monthly balances, number and amount of purchases, cash advance transactions, payments as well as customer demographics, and account details. Purchase transactions in particular were recorded per merchant code, tracking the purchasing habits of each customer. In addition, a multilevel grouping hierarchy of merchant codes was available in a relevant lookup table, enabling purchases per merchant category/subcategory to be summarized.

One of the things that had to be decided before building the mining dataset was the categorization level of the merchants. Since the bank's primary objective was to differentiate customers according to their general purchasing behaviors, a high-level categorization into 14 merchant categories (to be presented below) was selected. The use of a more detailed grouping was beyond the scope for the particular project and was left to a micro-clustering approach in a later phase. The preparation of the modeling dataset included the summarization of purchase transactions per merchant group.

Segmentation level was another issue that had to be determined. The selected data model and the record level of the final input table should be the same as the selected segmentation level. So, the marketers had to decide whether they would group together either customers or credit cards.

In general, each credit card customer can have more than one credit card account and each credit card account can include one primary and a number of potential add-on cards. In order to get a unified view of each customer, which would then facilitate their subsequent marketing strategies, the marketers decided to proceed with a customer level segmentation. Therefore they had to build the mining dataset at the customer level.

It had also been decided to use six months of data for the segmentation. A snapshot of the transactional data based on only a few days or even weeks might just capture a random twist in customer behavior. Since the objective was to develop a segmentation scheme that would reflect typical behaviors, the approach followed was to use a time frame (six months) large enough to avoid the pitfall of inconsistent usage patterns but not too large to take into account outdated behaviors.

SELECTING THE SEGMENTATION POPULATION

The segmentation project's target population initially included all the consumer customers. Initial data exploration, though, revealed that many of the bank's cards were inactive during the period examined. There was no spending at all during the previous six months, no purchases, and no cash advances. This finding was taken into account in the definition of the segmentation population.

Customers without any active cards and without spending at all were discarded from the input dataset. They comprised a segment of their own, an inactive segment which required special handling. Since inactivity is usually the first step toward termination of the relationship with an organization, a relevant list of all inactive customers was constructed and served as the basis for execution of a "reactivation" campaign. In this campaign, incentives were offered to these customers to tempt them to start using their cards again.

An examination of the balances revealed a special group of inactive customers for which the main reason for inactivity was reaching their entire credit limit. These customers carried past balances which they had not managed to pay and therefore they were not allowed any new spending. The approach for these customers was different. It did not include purchase incentives but an increase in their limit, at least for those who had shown good payment behavior (credit score) in the past.

The team involved in the project also decided to further narrow the segmentation population and to include only those customers who were systematically making purchases. Thus they imposed a business rule which filtered out those customers with a monthly average purchase amount below $50.

The final segmentation population consisted of:

- consumer customers, active at the time of the analysis, that is, customers with at least one open and active credit card (churned and inactive customers, namely customers whose cards were all closed or inactive, were excluded); and
- customers with monthly average purchases of at least $50 during the last six months.

The procedure for determining the final segmentation population is presented in Figure 6.1.

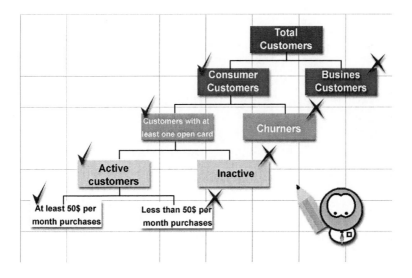

Figure 6.1   Selecting the population for credit card segmentation.

Some Useful KPIs of Credit Card Usage

A good approach for identifying customers who could benefit from a potential rise in their credit limit is to monitor the limit usage. A relevant indicator would measure the balance to credit limit ratio.

   Another useful indicator which measures the diversity of card usage and its adoption in everyday life is the number of different merchant categories with purchases.

   Other measures which can also be considered as core indicators for monitoring credit card usage include:

- Frequency of purchases and cash advance
- Recency of last purchase (time since the last purchase)
- Monthly average number and amount of purchase and cash advance transactions
- Monthly average balances
- Average purchase amount per transaction
- The percentage of total card spending attributable to the cash advance
- The payment to minimum payment ratio and the payment to statement balance ratio.

Table 6.1 summarizes the segmentation implementation design.

Table 6.1   Credit card segmentation: summary of implementation methodology.

Critical point	Decision
Segmentation level	Customer level
Segmentation population	Active customers with at least one open and active credit card and with monthly average purchases of at least $50 during the last six months
Data sources	Mining data mart and MCIF
Time frame of used data	Last six months
Segmentation fields	Purchase KPIs contained in the MCIF (percentages, averages, etc.) summarizing purchases of the last six months
Profiling fields	All fields of interest: usage KPIs of the last six months, demographics, etc.
Technique used	Clustering technique to reveal segments

THE SEGMENTATION FIELDS

The list in Table 6.2 presents the clustering inputs which were used for model training. It also presents the profiling fields that did not participate in the model building but only in the profiling of the revealed clusters.

Table 6.2   Fields used for the segmentation of credit card holders.

Field name	Description
Purchases by merchant category – clustering fields:	
PRC_ACCESSORIES	The percentage of total purchase amount spent at accessory stores (includes shoes/jewelry/cosmetics)
PRC_APPLIANCES	The percentage of total purchase amount spent at electrical appliances/electronics stores
PRC_CULTURE	The percentage of total purchase amount spent at cinemas/theaters/museums etc.
PRC_GAS	The percentage of total purchase amount spent at gas stations
PRC_BOOKS_MUSIC	The percentage of total purchase amount spent at bookstores and music/DVD/video/computer games retailers
PRC_APPAREL	The percentage of total purchase amount spent at clothing stores

Table 6.2   (*continued*)

Field name	Description
PRC_FITNESS	The percentage of total purchase amount spent at gyms/sport clubs/fitness centers/sports and outdoors stores
PRC_EDUCATION	The percentage of total purchase amount spent at educational institutes
PRC_ENTERTAINMENT	The percentage of total purchase amount spent at restaurants/coffee houses/bars/clubs etc.
PRC_FOOD	The percentage of total purchase amount spent at supermarkets
PRC_HEALTH	The percentage of total purchase amount spent for health purposes (medical insurance/hospital fees/medical examinations/pharmacies etc.)
PRC_HOME_GARDEN	The percentage of total purchase amount spent at home/garden and DIY stores
PRC_TELCOS	The percentage of total purchase amount spent on telecommunication services (fixed and mobile telephony)
PRC_TRAVEL	The percentage of total purchase amount spent on travel expenses (travel agencies/air tickets/hotel bookings/car rentals etc.)
ACCESSORIES	Monthly average purchase amount spent at accessory stores (including shoes/jewelry/cosmetics)
APPLIANCES	Monthly average purchase amount spent at electrical appliances/electronics stores
CULTURE	Monthly average purchase amount spent at cinemas/theaters/museums etc.
GAS	Monthly average purchase amount spent at gas stations
BOOKS_MUSIC	Monthly average purchase amount spent at bookstores and music/DVD/video/computer games retailers
APPAREL	Monthly average purchase amount spent at clothing stores
FITNESS	Monthly average purchase amount spent at gyms/sport clubs/fitness centers/sports and outdoors stores
EDUCATION	Monthly average purchase amount spent at educational institutes
ENTERTAINMENT	Monthly average purchase amount spent at restaurants/coffee houses/bars/clubs
FOOD	Monthly average purchase amount spent at supermarkets
HEALTH	Monthly average purchase amount spent for health purposes (medical insurance/hospital fees/medical examinations/pharmacies etc.)

(*continued overleaf*)

Table 6.2   (*continued*)

Field name	Description
HOME_GARDEN	Monthly average purchase amount spent at home/garden and DIY stores
TELCOS	Monthly average purchase amount spent on telecommunication services (fixed and mobile telephony)
TRAVEL	Monthly average purchase amount spent on travel expenses (travel agencies/air tickets/hotel bookings/car rentals etc.)
MERCHANT_TYPES	Number of different merchant categories with purchases
Other usage fields – profiling fields:	
BALANCE	Monthly average balance (based on daily balance averages)
PURCHASES	Monthly average purchase amount
ONEOFF_PURCHASES	Monthly average one-off purchase amount
INSTALLMENT_ PURCHASES	Monthly average of installment purchases
CASH_ADVANCE	Monthly average amount of cash advance
PURCHASE_ FREQUENCY	Frequency of purchases (percentage of months with at least one purchase)
CASH_ADVANCE_ FREQUENCY	Frequency of cash advance
SPENDING _FREQUENCY	Frequency of total card spending (purchases and cash advance)
PRC_ONEOFF_ PURCHASES	Amount of one-off purchases as a percentage of total card spending
PRC_CASH_ADVANCE	Cash advances as a percentage of total card spending
PRC_INSTALLMENTS_ PURCHASES	Amount of installments as a percentage of total card spending
AVERAGE_PURCHASE_ TRX	Average amount per purchase transaction
AVERAGE_CASH_ ADVANCE_TRX	Average amount per cash advance transaction
MINIMUM_PAYMENT_ RATIO	The ratio of monthly average payments to the monthly average of the minimum payment amounts
BALANCE_PAYMENT_ RATIO	The ratio of monthly average payments to monthly average statement balances

Table 6.2   (*continued*)

Field name	Description
PRC_FULL_PAYMENT	Percentage of months with full payment of the due statement balance
PRC_ZERO_PAYMENT	Percentage of months with no payments at all of the due statement balance
LIMIT_USAGE	The ratio of average balance to credit limit.
PRC_INTERNATIONAL	Amount spent abroad as a percentage of total spending amount
PRC_INTERNET	Web purchases as a percentage of total purchase amount
Credit account information fields – profiling fields	
Customer socio-demographics – profiling fields	

THE ANALYTICAL PROCESS

The analytical process that was followed included three main steps.

1. Extraction of the segmentation dimensions by using a supervised data reduction technique.
2. Identification of the customer segments through the application of a clustering model.
3. Profiling of the segments.

In Figure 6.2 we can see the IBM SPSS Modeler stream developed for the needs of segmentation. The data preparation part is omitted in this screenshot. The first step involved the application of a PCA command (node) for the data reduction and identification of the underlying data dimensions. Customers were then scored by the generated PCA model (the "diamond" node) and, in the final step of the procedure, the generated PCA scores were loaded as inputs for training a TwoStep cluster model.

The procedure concluded with the exhaustive profiling and evaluation of the derived solution, before accepting it and adopting it into the organization's procedures. The whole process is presented in detail in the following sections.

Revealing the Segmentation Dimensions

The clustering inputs included the set of fields denoting the percentage of purchases per merchant category. These fields were used instead of the average

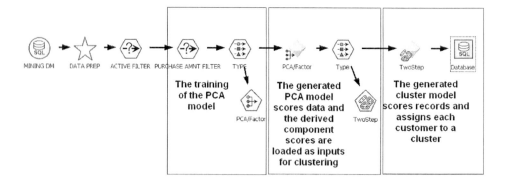

Figure 6.2   The IBM SPSS Modeler stream for clustering.

purchase amount per merchant since, after a few trials, it became evident that they produced a clearer separation which better reflected the relative spending preferences, adjusting for the differences between customers in terms of their overall spending amounts.

It is generally recommended that a data reduction technique, like principal components or factor analysis, be applied to the input fields before clustering. This approach was followed in the project presented here. The 14 fields of the percentage of purchases by merchant type were used as inputs to a PCA model which extracted a set of eight components, based on the "Eigenvalues over 1" criterion and a Varimax rotation. These eight components accounted for about 70% of the information carried by the original inputs, as shown in the "variance explained" table, Table 6.3.

Table 6.3   Deciding the number of extracted components by examining the "variance explained" table.

Total variance explained			
Components	**Eigenvalue**	**Percentage of variance**	**Cumulative %**
1	1.52	10.86	10.86
2	1.34	9.57	20.43
3	1.31	9.36	29.79
4	1.25	8.93	38.71
5	1.19	8.50	47.21
6	1.12	8.00	55.21
7	1.04	7.43	62.64
8	1.02	7.29	69.93
9	0.95	6.79	76.71

Table 6.3 (*continued*)

Total variance explained

Components	Eigenvalue	Percentage of variance	Cumulative %
10	0.84	6.00	82.71
11	0.73	5.21	87.93
12	0.70	5.00	92.93
13	0.59	4.21	97.14
14	0.40	2.86	100.00

As the 14 original fields represented a relatively high categorization level of merchants, the benefits of PCA in terms of data reduction may not be evident in this particular case. However, having to analyze tens or even hundreds of fields is not unusual in data mining. In these situations a data reduction technique can provide valuable help in understanding and "tidying" up the available information.

But what do these eight derived components represent? Since they were about to be used as the segmentation dimensions, it was crucial for the project team to understand them. The components were interpreted in terms of their association with the original fields. These associations (correlations or loadings) are listed "in the rotated component matrix" given in Table 6.4.

Table 6.4 Understanding and labeling the components through the rotated component matrix.

	Rotated component matrix							
	Components							
	1	2	3	4	5	6	7	8
PRC APPAREL	0.73							
PRC ACCESSORIES	0.66							
PRC FOOD		0.74						
PRC GAS		0.53						
PRC APPLIANCES			−0.86					
PRC TELCOS				−0.88				
PRC HEALTH					−0.93			
PRC HOME GARDEN						0.91		
PRC TRAVEL			0.44		0.32		0.82	
PRC FITNESS							0.31	

(*continued overleaf*)

Table 6.4   *(continued)*

| | Rotated component matrix | | | | | | | |
| | Components | | | | | | | |
	1	2	3	4	5	6	7	8
PRC ENTERTAINMENT						−0.36	0.31	0.55
PRC EDUCATION								0.51
PRC CULTURE								0.50
PRC BOOKS MUSIC								0.36

Trivial correlations (below 0.3 in absolute value) are omitted. As a reminder, we note that values close to 1, in absolute value, signify strong correlations and original fields that are associated with a specific component. The interpretation and labeling of the derived components is summarized in Table 6.5.

Table 6.5   Interpretation of the extracted principal components.

Derived Components	
Component	**Label and Description**
1.	**Fashion purchases:** Component 1 represents fashion purchases. Purchase of clothes and accessories are correlated with each other and with component 1. Therefore customers with high spending at these merchants will also have a high (positive) component 1 score
2.	**Daily needs spending:** Component 2 seems to measure the relative card usage for purchases at supermarkets and gas stations
3.	**Appliance purchases:** The next component is associated with purchases of appliances. Travel spending is also (moderately) negatively correlated with this component and with appliance purchases: not many appliance buyers spent money on traveling. Hence this component captures this contrast: travelers are expected to have a relatively high positive score on component 3, whereas appliance buyers have a large negative score
4.	**Spending on telecommunication services:** Component 4 is associated with using the card for telecommunication services. Large negative scores suggest increased relevant usage

Table 6.5   (*continued*)

Derived Components	
Component	**Label and Description**
5.	**Health services:** Purchases for health services are loaded high on this component
6.	**Home/garden and DIY store purchases:** The next component measures the usage in home/garden/DIY stores
7.	**Travel spending:** Component 7 represents use of the card to cover traveling expenses
8.	**Entertainment/education:** Spending on entertainment, culture, and education services load high on this component

The PCA model generated the relevant component scores which were subsequently used as clustering inputs in the next phase of the project. These scores are standardized. Scores above or below the overall mean value of 0 indicate groups which deviate from the "average" behavior (spending more or less than the average).

Identification and Profiling of Segments

The extracted component scores were subsequently used as inputs in a TwoStep clustering algorithm to reveal groups of customers with similar purchasing habits. The automatic clustering procedure of the algorithm suggested an eight-cluster solution (Figure 6.3). The quality of the clustering solution was good with an overall silhouette measure of 0.6.

The distribution of the revealed clusters is shown in Table 6.6. Clusters 1 and 6 were the larger ones, jointly accounting for almost half of the credit card holders.

At this point, the main task of the project team was to understand and label the derived groups and assess their business significance and value. This process involved the application of simple reporting techniques to examine the structure of each cluster in terms of the PCA scores and the original usage attributes in order to reveal the differentiating characteristics of each cluster.

The PASW Cluster Viewer was used to examine the grid of cluster centers and the mean values of the component scores for each cluster (Figure 6.4). Since the component scores are standardized measures, their overall population means are 0. Therefore positive values indicate scores above the average and

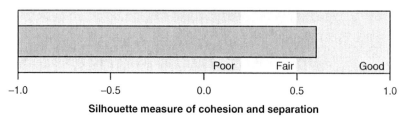

Model Summary

Algorithm	TwoStep
Input Features	8
Clusters	8

Cluster Quality

Poor   Fair   Good

−1.0   −0.5   0.0   0.5   1.0

Silhouette measure of cohesion and separation

Figure 6.3 The overall silhouette measure of the clustering solution for credit card holders.

Table 6.6 Distribution of the derived clusters.

		Percentage of credit card holders
Clusters	Cluster 1	21.3
	Cluster 2	10.0
	Cluster 3	11.4
	Cluster 4	9.1
	Cluster 5	10.0
	Cluster 6	25.4
	Cluster 7	5.0
	Cluster 8	7.7
Total		100.0

increased spending, while negative values denote scores below the average and lower spending. In this particular example, though, we have to bear in mind that negative scores in components 3, 4, and 5 represent increased spending for appliances, telecommunication, and health services respectively, due to the high negative loadings of the original inputs with those components.

Additionally, the clusters were profiled in terms of the original purchase fields. The two affinity tables that follow, Tables 6.7 and 6.8, summarize the percentage and the absolute purchase amount of each cluster by merchant type.

Clusters

Feature
Importance
☐1

Cluster Label	cluster-1	cluster-2	cluster-3	cluster-4	cluster-5	cluster-6	cluster-7	cluster-8
Description								
Features	Component_1 – Fashion −0.16	Component_1 – Fashion −0.67	Component_1 – Fashion −0.45	Component_1 – Fashion −0.56	Component_1 – Fashion −0.60	Component_1 – Fashion 1.77	Component_1 – Fashion −1.16	Component_1 – Fashion −0.74
	Component_2 – Daily needs 0.02	Component_2 – Daily needs −0.54	Component_2 – Daily needs 2.76	Component_2 – Daily needs −0.68	Component_2 – Daily needs −0.42	Component_2 – Daily needs −0.26	Component_2 – Daily needs −1.03	Component_2 – Daily needs −0.70
	Component_3 – Appliances −0.05	Component_3 – Appliances 0.36	Component_3 – Appliances 0.17	Component_3 – Appliances 1.44	Component_3 – Appliances 0.30	Component_3 – Appliances 0.08	Component_3 – Appliances 1.03	Component_3 – Appliances −3.36
	Component_4 – Telcos −0.02	Component_4 – Telcos −2.96	Component_4 – Telcos 0.23	Component_4 – Telcos 0.90	Component_4 – Telcos 0.44	Component_4 – Telcos 0.14	Component_4 – Telcos 1.53	Component_4 – Telcos 0.58
	Component_5 – Health −0.04	Component_5 – Health 0.37	Component_5 – Health 0.16	Component_5 – Health 1.10	Component_5 – Health −2.12	Component_5 – Health 0.07	Component_5 – Health 0.55	Component_5 – Health 0.47
	Component_6 – Home/Garden −0.01	Component_6 – Home/Garden −0.40	Component_6 – Home/Garden −0.18	Component_6 – Home/Garden −0.08	Component_6 – Home/Garden 2.01	Component_6 – Home/Garden −0.13	Component_6 – Home/Garden −1.47	Component_6 – Home/Garden −0.31
	Component_7 – Travel −0.06	Component_7 – Travel −0.20	Component_7 – Travel −0.19	Component_7 – Travel 1.22	Component_7 – Travel −0.12	Component_7 – Travel −0.07	Component_7 – Travel 0.05	Component_7 – Travel 0.01

Figure 6.4   The table of cluster centers.

The cluster centroids and these tables convey similar results. Most of the clusters seem to be related to specific merchant categories and purchasing habits. More specifically, cluster 1 contains customers with average spending in many different merchant categories. These customers have the most diverse purchasing habits. They present moderate usage and tend to buy a variety of different products with their cards. On the other hand, the rest of the clusters have more partial preferences and seem more tightly connected with the buying of specific products. Cluster 2 is associated with fees for telecommunication services and cluster 3 is characterized by increased spending at supermarkets and gas stations. Moreover, cluster 4 shows increased usage for traveling expenses. Spending at home/garden stores and for health purposes appears to be a defining characteristic of cluster 5. Customers assigned to cluster 6 tend to buy clothing and accessories. Cluster 7 relates to spending for cultural/educational reasons and entertainment, whereas customers in cluster 8 use their cards mainly for the purchase of appliances.

It is always a good idea to depict results in a graphical way so that the profiles of each cluster are clearly illustrated. The bubble plot in Figure 6.5 is based on aggregated data. Each "dot" represents a cluster. The X-axis co-ordinate of each dot or cluster denotes its mean percentage of purchases at apparel stores, the

Table 6.7   Mean percentage of purchases by merchant type for each cluster.

	Clusters								
	Cluster 1	Cluster 2	Cluster 3	Cluster 4	Cluster 5	Cluster 6	Cluster 7	Cluster 8	Total
PRC ACCESSORIES (%)	7.2	2.3	2.7	3.4	2.9	23.5	3.5	1.2	9.0
PRC APPLIANCES (%)	13.5	5.3	3.4	4.7	3.8	2.8	4.7	90.3	12.5
PRC CULTURE (%)	0.2	0.1	0.0	0.0	0.1	0.1	3.4	0.0	0.3
PRC GAS (%)	5.7	1.5	28.4	3.5	1.5	0.4	2.2	0.8	5.4
PRC BOOKS MUSIC (%)	1.8	0.4	0.5	0.6	0.3	0.5	14.1	0.2	1.4
PRC APPAREL (%)	27.4	7.2	5.7	12.7	9.2	63.0	9.1	3.4	26.0
PRC FITNESS (%)	0.3	0.1	0.0	0.2	0.0	0.0	2.5	0.0	0.2
PRC EDUCATION (%)	0.0	0.0	0.0	0.4	0.0	0.0	4.2	0.0	0.3
PRC ENTERTAINMENT (%)	3.2	0.7	0.5	2.9	0.6	0.5	44.2	0.4	3.5
PRC FOOD (%)	18.8	2.2	54.4	2.8	2.4	2.8	2.4	1.1	11.6
PRC HEALTH (%)	3.4	1.0	0.8	1.3	38.5	1.8	1.6	0.5	5.5
PRC HOME GARDEN (%)	7.4	1.6	1.2	2.4	37.1	1.8	1.9	0.8	6.4
PRC TELCOS (%)	5.6	74.9	1.0	3.1	1.6	0.9	2.7	1.0	9.8
PRC TRAVEL (%)	5.5	2.7	1.3	62.1	1.9	1.8	3.6	0.4	8.2

Table 6.8 Mean purchase amount by merchant type for each cluster.

					Clusters				
	Cluster 1	Cluster 2	Cluster 3	Cluster 4	Cluster 5	Cluster 6	Cluster 7	Cluster 8	Total
ACCESSORIES	13.8	3.3	4.2	8.1	4.4	30.1	4.7	1.0	**12.9**
APPLIANCES	16.1	6.5	5.0	9.3	4.6	3.4	6.3	71.0	**12.6**
CULTURE	0.5	0.1	0.1	0.2	0.1	0.3	4.9	0.0	**0.5**
GAS	9.1	2.2	32.1	8.4	3.1	.7	5.1	1.0	**7.4**
BOOKS/MUSIC	3.2	0.6	1.2	1.7	0.5	0.8	13.2	0.2	**2.0**
APPAREL	44.5	10.6	9.8	29.6	14.2	71.8	13.0	2.7	**34.9**
FITNESS	0.2	0.1	0.1	0.4	0.0	0.0	4.3	0.0	**0.3**
EDUCATION	0.3	0.0	0.1	1.2	0.1	0.2	16.2	0.0	**1.0**
ENTERTAINMENT	7.0	1.2	1.2	9.3	1.3	1.2	113.0	0.3	**8.7**
FOOD	37.1	3.0	85.8	6.3	4.1	3.6	3.7	1.2	**20.2**
HEALTH	7.1	1.4	1.7	3.5	61.2	3.0	2.8	0.4	**9.2**
HOME/GARDEN	11.5	2.2	2.6	7.1	37.5	3.3	3.6	0.7	**8.5**
TELCOS	10.9	82.7	2.4	8.1	3.2	1.7	5.4	1.1	**12.7**
TRAVEL	14.2	5.2	3.5	125.0	4.1	4.6	9.3	0.5	**17.4**

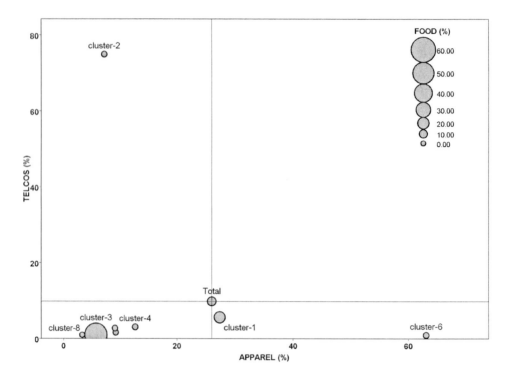

Figure 6.5 Bubble plot illustrating the percentage of purchases at supermarkets, clothing stores, and telecommunication operators per cluster.

Y-axis co-ordinate the spending on telecommunication services, and the dot size the spending at supermarkets. The reference lines correspond to the overall means and hence represent the "average" credit card holder.

The bubble plot outlines the relation of cluster 6 to purchases of clothes/accessories and the association of cluster 2 with telecommunication services. Their percentage of purchases at relevant merchants is by far the largest of all clusters. The larger size of the dot corresponding to cluster 3 designates the high spending of this cluster at supermarkets. The remaining clusters are not clearly differentiated by the three usage fields summarized in this plot.

An analogous bubble chart is presented in Figure 6.6, summarizing the derived clusters in terms of relative spending on entertainment, traveling, and home/garden products.

Cluster 5 has the highest percentage of spending at home stores. Cluster 7 is associated with entertainment and cluster 4 with travel expenses. In both bubble plots, cluster 1 is close to the reference lines and shows some proximity to the average profile.

The next step, after the exploration of the relative spending by merchant categories, was the examination of possible associations of the clusters with other

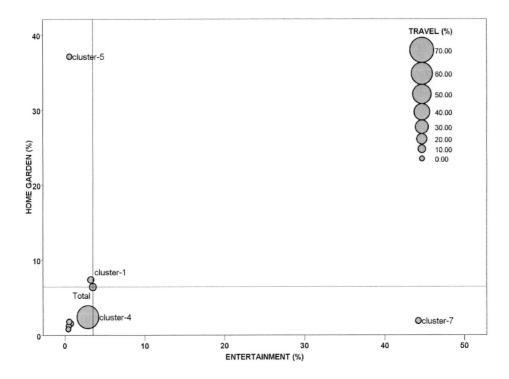

Figure 6.6   Bubble-plot illustrating the percentage of purchases at home stores, for traveling, and for entertainment per cluster.

attributes which had not participated in the building of the cluster model, including general usage fields and demographic information.

The separation of the clusters is reflected in the main usage KPIs presented in Table 6.9. Cluster 8 has the lowest purchase amount as opposed to clusters 4, 7, and 1. Clusters 3, 6, and 7 carry the highest balances. The international usage of the cards is generally low, except by clusters 4 and 7. Those two clusters also have the highest e-shopping usage. With respect to their demographic profile (Table 6.10), clusters 2, 4, 6, and 7 concern a large number of younger customers as opposed to cluster 5 which seems to contain mostly mature people. Women form the majority in cluster 6 and men in cluster 3. Clusters 3 and 5 are mainly composed of married customers. The smaller percentages of married customers can be found in clusters 2, 4, and 7.

The profiling phase concluded with the application of statistical hypothesis tests on the percentage purchases fields. These tests compared the mean values of each cluster to the overall population means and identified statistically significant differences. The test results have the format of profiling charts. The direction of the bars indicates the relation between the cluster and the overall population. The

Table 6.9 Usage KPIs by cluster.

| | Clusters | | | | | | | |
	Cluster 1	Cluster 2	Cluster 3	Cluster 4	Cluster 5	Cluster 6	Cluster 7	Cluster 8	Total
PURCHASES	175	119	150	218	139	125	205	80	148
BALANCE	2230	1963	2507	1900	2053	2388	2402	2050	2222
AVERAGE PURCHASE TAX	108	84	117	183	264	103	162	164	135
PURCHASE FREQUENCY (%)	78	75	83	66	61	67	63	41	69
PRC INTERNATIONAL (%)	5	4	2	25	1	2	10	1	5
PRC INTERNET (%)	8	5	2	15	1	8	29	6	8

Table 6.10 The demographical profile of each cluster.

		Cluster 1	Cluster 2	Cluster 3	Cluster 4	Cluster 5	Cluster 6	Cluster 7	Cluster 8	Total
Gender (%)	Male	45	55	62	48	43	28	47	45	44
	Female	55	45	38	52	57	72	53	55	56
Age (%)	18–24	17	18	5	25	1	21	35	16	17
	25–34	25	35	15	34	3	35	29	27	26
	35–44	30	30	45	35	15	35	20	32	32
	45–54	18	15	29	5	40	8	10	17	17
	55+	10	2	6	1	41	1	6	8	8
Marital (%)	Married	65	55	95	55	75	62	52	67	66
	Single/divorced	35	45	5	45	25	38	48	33	34

relevant dotted reference lines correspond to the test's critical value and designate whether the observed differences can be considered statistically significant. Thus, bars on the right (left) part of the Y-axis designate mean cluster values higher (lower) than the total mean. Furthermore, a bar surpassing the reference line indicates that the observed difference cannot be attributed to chance alone. These charts reveal the defining characteristics of the clusters and supplement the full profile of each cluster that follows.

Segment 1 – Typical Users (Figure 6.7)

Behavioral Profile

- Diverse card usage with purchases at different merchants
- Relatively increased total (monthly average) purchase amount
- Increased purchase frequency
- Moderate (monthly average) balance amount.

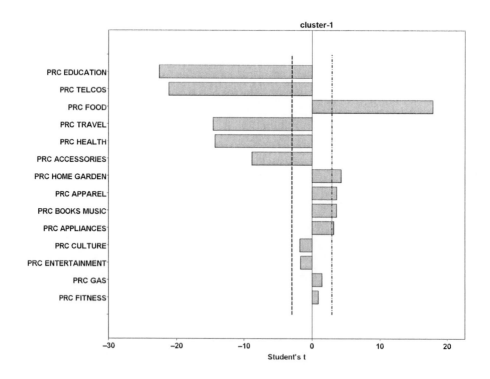

Figure 6.7   Cluster 1 profiling chart.

Segment Size

21%

Contribution to Overall Purchase Amount

25%

Demographic Profile

Typical demographic profile.

Segment 2 – Telcos Fans (Figure 6.8)

Behavioral Profile

- Use of card mainly for payment of telecommunication services
- Moderate total purchase amount with increased frequency (perhaps due to payment of telecommunication services at a regular basis) of card usage
- Relatively low balances.

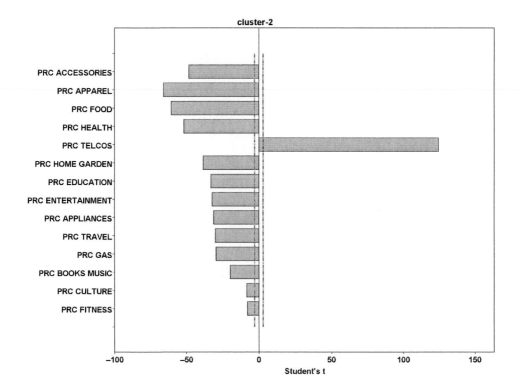

Figure 6.8   Cluster 2 profiling chart.

Segment Size

10%

Contribution to Overall Purchase Amount

8%

Demographic Profile

Predominantly young and single cardholders.

Segment 3 – Family Men (Figure 6.9)

Behavioral Profile

- Use of the card to cover the daily needs
- Use of the card mainly at supermarkets

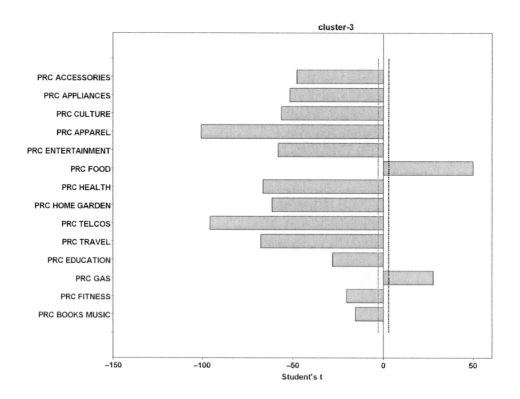

Figure 6.9   Cluster 3 profiling chart.

- Also use of the card at gas stations
- The highest balances
- Frequent purchase transactions but of moderate amounts.

Segment Size

11%

Contribution to Overall Purchase Amount

12%

Demographic Profile

Almost exclusively comprises married people. Predominantly male customers over 35 years old.

Segment 4 – Travelers (Figure 6.10)

Behavioral Profile

- Use of card mainly for traveling and vacations (hotel bookings/air tickets/car rentals, etc.)
- The highest international usage with a large percentage of the total purchases made abroad
- Increased purchases over the Internet (for e-booking?)
- Increased total amount of purchases (tickets are expensive!)
- Infrequent but high-value purchase transactions
- Relatively low balances.

Segment Size

9%

Contribution to Overall Purchase Amount

13%

Demographic Profile

Predominantly unmarried and young card holders.

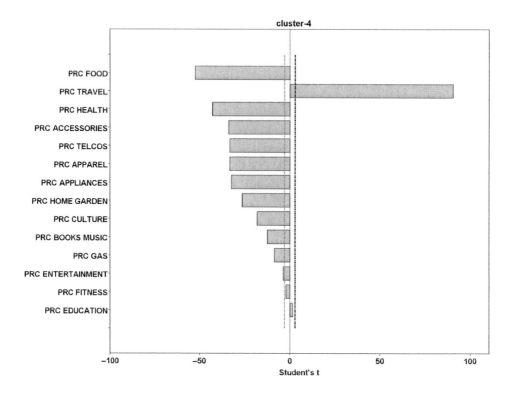

Figure 6.10 Cluster 4 profiling chart.

Segment 5 – Oldies (Figure 6.11)

Behavioral Profile

- Use of the card mainly at home/garden/DIY stores
- Use of the card for health purposes (medical insurance, hospital fees, medical examinations, pharmacies, etc.)
- The highest average purchase amount per transaction
- Infrequent but high-value purchase transactions
- The lowest international and Internet usage.

Segment Size

10%

Contribution to Overall Purchase Amount

9%

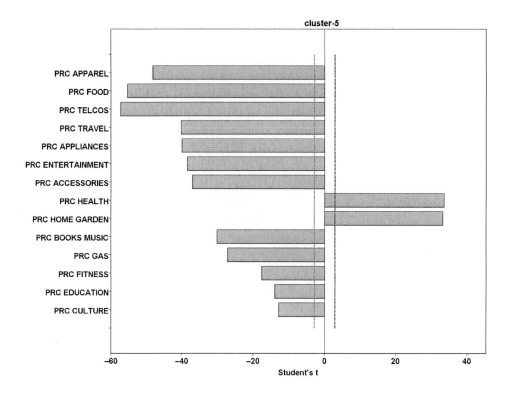

Figure 6.11   Cluster 5 profiling chart.

Demographic Profile

Mostly married and mature users.

Segment 6 – Fashion Ladies (Figure 6.12)

Behavioral Profile

- Increased purchases at apparel stores
- Increased purchases at accessory stores (jewelry, shoes, etc.)
- Increased balances.

Segment Size

25%

Contribution to Overall Purchase Amount

21%

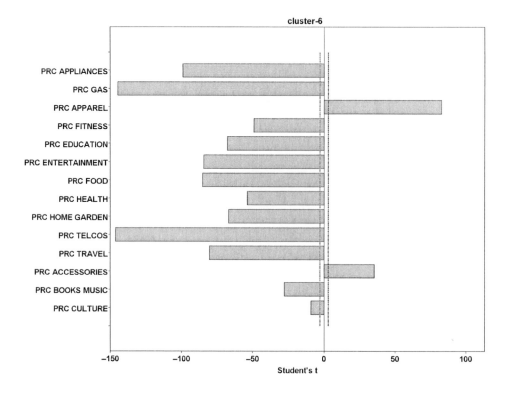

Figure 6.12   Cluster 6 profiling chart.

Demographic Profile

Consists of young/middle-aged women.

Segment 7 – Fun Lovers (Figure 6.13)

Behavioral Profile

- Use of the card for entertainment (restaurants, coffee houses, etc.)
- Increased purchase amount spent at bookstores and music/DVD/video/computer games retailers
- Also use of the card for paying educational fees
- Use of the card at gyms/sport clubs/fitness centers/sports and outdoors stores
- Increased total purchase amount
- Increased balances
- Increased international and the highest Internet usage.

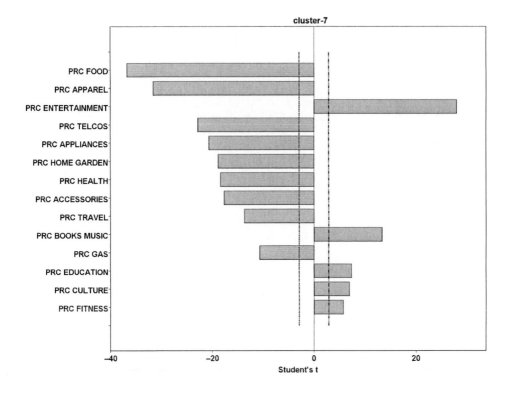

Figure 6.13   Cluster 7 profiling chart.

Segment Size

5%

Contribution to Overall Purchase Amount

7%

Demographic Profile

Mostly young unmarried customers.

Segment 8 – Occasional Users (Figure 6.14)

Behavioral Profile

- Mainly spending for the purchase of electrical appliances and electronics
- Lowest purchase frequency
- Infrequent but high-value purchase transactions
- Low total purchase amount and balances.

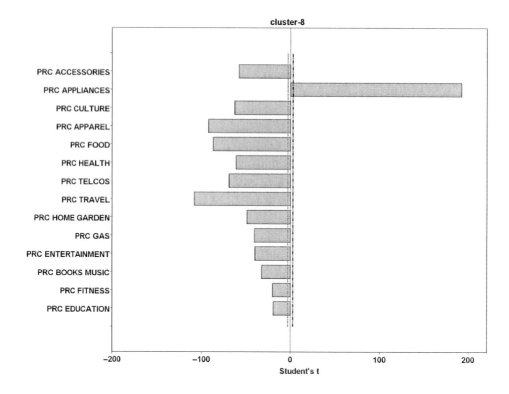

Figure 6.14 Cluster 8 profiling chart.

Segment Size

8%

Contribution to Overall Purchase Amount

4%

Demographic Profile

Typical demographic profile.

Investigating Segment Structure with Supervised Models

The profiling of the segments can be enriched by using supervised models, decision trees in particular, trained with the segment membership field as target and the key profiling fields as inputs. Decision trees can provide a multi-attribute profile of the segments and discern their defining characteristics. Apart from giving some insight into the clusters, the decision trees'

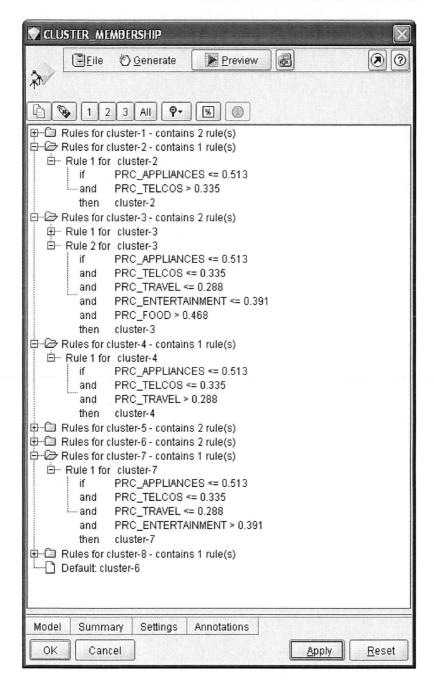

Figure 6.15   Using a decision tree to profile segments.

classification rules can be used for scoring and the assignment of new records to segments.

Another useful approach is the development of a decision tree using as inputs only those fields that are available upon customer registration, for instance demographic information and contract/account details. This proactive model can then be used immediately after the registration process of new customers to estimate their future behavior and assign them to the appropriate segments. Obviously this type of assignment provides a short-term, early-stage estimation and should be replaced with the actual behavioral segmentation scheme for established customers.

Figure 6.15 presents a subset of the decision tree rules derived for the credit card holder segments. The model was trained using the set of fields denoting the percentage of purchases by merchant categories as inputs. Note how the decision tree has identified the profiling rules for each of the eight segments. In the displayed set of rules, we can see the association of segment 2 with telecommunication services, while segment 3 is associated with supermarket purchases, segment 4 with use for traveling, and cluster 7 with spending on entertainment.

USING THE SEGMENTATION RESULTS

Once the segmentation scheme was finalized and approved, it was then deployed. An update process was designed and implemented. It was scheduled to run on a monthly basis in order to keep the segmentation scheme informed and to monitor the possible movements of credit card holders across segments. The deployment procedure ensured the storage and distribution of segment information to decision makers and authorized marketing users and the loading of the segments onto the operational CRM systems.

One evident conclusion of the segmentation analysis was that most of the existing customers seemed to be using their cards for specific purposes. The majority of credit card holders were characterized by non-diverse usage, a sign of a non-mature market. Only "Typical users" seemed to use their cards to cover various types of needs. Thus, a first marketing action that was decided was to motivate customers in the remaining segments to start using their cards in all aspects of their everyday life. To this end, a marketing campaign was set up, which offered incentives to customers in order to encourage them to expand and diversify their card usage.

The derived scheme identified different customer typologies and revealed different needs and preferences of the customer base. This information was then used as a guide for the development of **new specialized products/reward programs and tailored offers/incentives** in order to achieve more specialized customer handling.

As for the offers and incentives granted to customers, the adopted approach included their adjustment to the distinct customer profiles portrayed by the revealed segments. For instance, the bank encloses promotional brochures detailing special prices and deductions for its customers in the monthly mailing of their balance statements. This promotional material was tailored to each segment's purchase profile, instead of flooding all the customers with every offer. Similarly, the bank's up-selling and churn prevention campaigns include the offer of incentives for promoting usage and inhibiting attrition, respectively. These incentives were adapted to each segment's particular characteristics. For instance, a member of the "Travelers" segment should be included in a draw for air tickets, or a member of the "Fun lovers" segment in a draw for concert tickets, provided they increased their card spending.

The segmentation results were also used in new product development and the offer of customized products and reward programs. The selected strategy involved offering specialized types of cards, with suitable rewards covering the differentiated needs of customers.

"Telcos fans," for instance, appeared to use their card mainly to pay for telecommunication services. Thus, the adopted marketing strategy included the bank's partnership with a telephone operator in the development of a new "telcos" card. This card bore the operator's logo and could also be issued at the provider's points of sale. It incorporated a reward scheme that offered reduced telecommunication fees or free call minutes if the total card purchases exceeded a specific amount.

For the "Family men" of segment 3, the bank's proposition included a card with rebates and gift certificates based on card usage at gas stations and supermarkets.

A travel-oriented card was the bank's offer to the "Travelers" segment. Partnerships with specific airlines, travel agencies, hotels, and car rental companies made it possible to offer specific benefits to frequent travelers. For instance, customers could take advantage of the card's reward program and use their cards to collect air miles or get a special booking discount at a partnering hotel. The card's benefits also included the offer of travel insurance when traveling abroad. Additionally, the bank started to address these customers through its subsidiary travel agency and to send them travel offers on a regular basis. All customers began to receive a leaflet on travel bargains and the card was further advertised at airports, on travel Internet sites, and in magazines.

A new card was also developed for the "Mature users" segment. Benefits of this card included zero-rate installments and discounts on purchases at partnering home stores. Moreover, as members of this segment had shown high spending on health expenses, the bank also partnered with an insurance company to include health benefits in the card's advantages, such as increased credit limit for health expenses.

A "fashion" card with a relevant reward scheme was the bank's proposition for the "Fashion ladies" of segment 6. The rewards tied to card usage included rebates and zero-rate installments for purchases at retail affiliates. This card was advertised in women's fashion magazines. Stands with card brochures were also positioned in affiliated fashion stores. A bargain fashion catalogue was also mailed to members of this segment, promoting card purchases at partnering stores.

"Fun lovers" seemed perfect candidates for a "student" card. The rewards of this card were tied to card usage at restaurants, bars, clubs, cinemas, bookstores, and so on. The card was advertised in magazines, on TV shows, and on web sites that were popular among young people, the advertised message presenting a modern and cool brand image.

A final type of card, designed and developed to target the "Occasional users" of segment 8, offered reductions and zero-rate installments when used at partnering retailers of electrical appliances and electronics. This card could also be issued by these authorized retailers on behalf of the bank.

BEHAVIORAL SEGMENTATION REVISITED: SEGMENTATION ACCORDING TO ALL ASPECTS OF CARD USAGE

After segmenting customers according to their purchasing habits and the products/services they spend their money on, the bank also realized the need for a second segmentation which would focus on general usage aspects such as frequency and intensity of spending, type of spending (purchase or cash advance), and payment behavior.

Although the bank could have used all the available purchase and usage data in an integrated segmentation model aiming to cover all behavioral aspects, it turned out that implementing two separate and more targeted segmentations provided a clearer separation. In the end, the two distinct segmentation schemes were combined in order to provide a unified profile of each customer in terms of both general usage and purchasing habits.

The general usage segmentation encompassed all the usage fields listed in Table 6.2. Once again the input dataset covered a time period of six months. The segmentation population included all active credit card holders, regardless of their monthly purchase amounts. The basic data dimensions were uncovered using a principal components model before the application of cluster analysis. To make a long story short, the main steps of the procedure were analogous to the ones presented in the previous project. So we will skip these steps and proceed directly to the derived results and the profiling of the revealed clusters.

The selected segmentation scheme was composed of six customer groups. Table 6.11 presents the means of some key usage indicators by cluster.

Table 6.11 Usage KPIs by cluster.

| | Clusters | | | | | | |
	Cluster 1	Cluster 2	Cluster 3	Cluster 4	Cluster 5	Cluster 6	Total
BALANCE	2078	4132	1527	850	92	1136	1652
PURCHASES	123	86	236	80	32	133	119
ONEOFF PURCHASES	95	50	165	6	9	131	77
INSTALLMENT PURCHASES	28	36	70	74	23	2	42
CASH ADVANCE	34	424	16	6	22	4	79
PURCHASE FREQUENCY	0.18	0.42	0.91	0.85	0.37	0.31	0.53
CASH ADVANCE FREQUENCY	0.06	0.79	0.05	0.02	0.03	0.02	0.15
PRC ONEOFF PURCHASES	0.61	0.08	0.63	0.04	0.18	0.93	0.40
PRC CASH ADVANCE	0.22	0.85	0.08	0.04	0.20	0.04	0.23
PRC INSTALLMENTS PURCHASES	0.17	0.07	0.30	0.91	0.62	0.03	0.37
AVERAGE PURCHASE TRX	31	53	73	56	68	229	74
LIMIT USAGE	0.92	0.87	0.32	0.35	0.04	0.37	0.50
PRC FULL PAYMENT	0.02	0.05	0.56	0.26	0.25	0.41	0.25
BALANCE PAYMENT RATIO	0.10	0.10	0.77	0.52	0.52	0.73	0.44

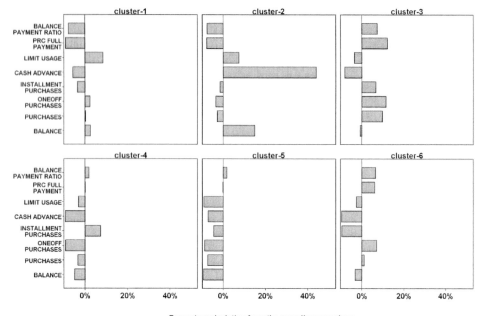

Figure 6.16 Usage KPIs by cluster.

The profile of the clusters is supplemented by the chart in Figure 6.16 which depicts percentage deviations from the overall mean values. Clusters were then labeled accordingly and concluding tables which summarized their differentiating characteristics were constructed, as follows.

Segment 1: Revolvers

Behavioral Profile

- Increased balances.
- Average purchases, mainly one-off.
- Low cash advance usage.
- Increased balances compared to their purchases suggest that they tend to carry past balances; this conclusion is also indicated by increased usage of their limit (balance to credit limit ratio).
- They seem to pay a small amount of their due balance every month, as indicated by the low payments to statement balance ratio. Their "revolver" behavior is also suggested by the fact that they have the lowest percentage of months with full payment of their due balance.

Segment Size

21%

Marketing Tips

They could take advantage of a possible increase in their credit limit, provided, of course, that they have good credit risk scores.

Segment 2: Cash Advance Users

Behavioral Profile

- Highest cash advance usage in terms of frequency and amount of cash withdrawals.
- Few purchases.
- 85% of their total card spending is attributed to cash advances.
- Highest monthly balance amounts.
- They also present high limit usage.
- Like "Revolvers," they tend to carry past balances and pay only a part of their monthly balances as indicated by the low frequency of full payments and the similarly low ratio of payments to balance.

Segment Size

15%

Marketing Tips

- New product development: a card with additional benefits for cash advance usage
- Possible prospects for personal loan products
- They could take advantage of a possible increase in their credit limit, provided, of course, that they have good credit risk scores.

Segment 3: Heavy Buyers – Transactors

Behavioral Profile

- Highest purchases, in terms of frequency and volume.
- One-off as well as installment purchases.
- Moderate balances and limit usage.
- Their spending is almost exclusively purchases since their cash advance usage is low.

- Their high payments to balance ratio and frequency of full payments suggest a transactor behavior.

Segment Size

20%

Marketing Tips

- Rewards and incentives (bonus points, rebates credited to their account) to motivate them to further increase their purchases; these rewards should be tailored to their purchase profile which was identified in the segmentation of purchasing habits.
- Candidates for product switch to the type of card that matches their purchasing preferences as revealed in the segmentation of purchasing habits.

Segment 4: Installment Buyers

Behavioral Profile

- Relatively low purchases.
- Their spending is almost exclusively on (old?) installments.
- Moderate balances.

Segment Size

18%

Marketing Tips

Incentives to start using their card for new spending. They could be included in a reactivation campaign.

Segment 5: Dormant Customers

Behavioral Profile

- Lowest purchases and cash advance usage.
- Lowest ratio of balance to credit limit.
- Insignificant (remnant?) balances.

Segment Size

15%

Marketing Tips

- Offering of incentives to start using their card.
- Since inactivity is usually the threshold of attrition, these customers should be given special attention, especially if they were high-value customers in the past.
- They may also hold competing credit cards and use them as their main cards; thus suitable candidates for a balance transfer offer and for inclusion in an "acquisition" campaign.

Segment 6: One-Off/Occasional Buyers

Behavioral Profile

- Increased purchases, including almost exclusively one-off purchases.
- High-value but infrequent purchase transactions: their purchase frequency is relatively low; however, they have the largest average amount per purchase transaction.
- Moderate balances and credit limit usage.
- They present a transactor behavior.

Segment Size

11%

Marketing Tips

Suitable candidates for product switch to the type of card that matches their purchasing preferences as revealed in the segmentation of purchasing habits.

THE CREDIT CARD CASE STUDY: A SUMMARY

The objective of a bank was to segment its consumer credit card holders according to their usage behaviors. The approach followed included a two-fold segmentation. The first segmentation grouped customers according to their purchasing habits and the product/services they tended to buy. The segmentation population included only active consumer customers with substantial purchases during the period examined. A high level of the merchant categorization scheme was selected for the analysis and customers were grouped according to their relative spending in these categories. The application of a clustering model revealed distinct customer groups with similar purchasing patterns. The bank's plans included using the segmentation information to provide customized offers and incentives and tailored card types that would better match the different habits of each customer group.

The second segmentation was based on more general behavioral criteria such as purchase frequency and volume, cash advance usage, and payment behavior. The segmentation covered all active consumer card holders and revealed six distinct groups, among which were a group of revolver customers, a group of cash advance users, and a group of heavy purchasers/transactors.

The two segmentation schemes were combined in the end to provide a complete view of customer behaviors.

SEGMENTATION IN RETAIL BANKING

There is no "average" customer. The customer base comprises distinct persons with differentiated needs that should be handled accordingly. In times when all banks seem to recognize the importance of customer experience and its effect on loyalty, segmentation is a principal strategy for improving relationships with existing customers and for luring customers from the competition.

Segmentation aims at the identification of homogeneous groups of people with dissimilar requirements that should be managed with customized strategies. A widely used segmentation scheme in retail banking divides customers according to their balances, product ownership, and total number of transactions into groups such as "Affluent," "Mass," "Inactive," and so on. These core segments, although established and undoubtedly useful in providing a first fundamental separation, fail to provide the necessary insight for the development of specialized management based on identified customer needs and behaviors. In order to design "personalized" products, rewards, and incentives, a more refined approach is needed that would further segregate the core segments and reveal the different customer typologies in terms of demographics/life-stage position, psychographics (attitudes/opinions/perceptions), profitability, and behavior. Different segmentation criteria correspond to different segmentation schemes which could be combined in such a way to help marketing in the design of tailored customer-centric strategies.

In the next sections we will focus on the efforts of a bank to develop an effective segmentation solution that would represent the different types of relationships with its consumer customers.

WHY SEGMENTATION?

Different people use different banking products and services for different reasons. However, they are all part of the bank's customer base and demand a handling that best suits their characteristics. Nowadays banks offer a large variety of products to cover these ever-changing demands and needs. The time when banks were places to simply keep your money safe is long gone. The relationship with the bank is not limited to just opening a plain savings account. Today the relationship typically includes many accounts and different products, often spread across different banks. Apart from deposit and savings accounts, banks also offer investment products, loans and mortgages, credit/debit cards, insurance products, and so on. All these options result in different product baskets and portfolios. Certain customers have a rather conservative profile and only maintain savings accounts. Others are willing to take a low or high risk and put a substantial share of their money into investment products, while for others the main relationship with the bank is a form of lending, a mortgage or a loan for example.

Customers are also diversified according to the number, volume, type, and preferred channel of the transactions they make. Visiting a branch and interacting with the bank's teller is no longer the only way to make banking transactions. Customers also have the option to use alternative channels, including ATMs, the Internet, phone, SMS, and so on, and to make different types of transactions such as deposits, withdrawals, payments, and so on. Some customers still insist on the traditional channels and the "human interaction" with bank employees, while others, appreciating the benefits of automated transactions, take advantage of alternative channels. Furthermore, certain customers have their salary account at a bank and therefore may select this bank as the main supplier of financial services.

In addition to all these behavioral/relationship sources of potential customer separation, we should not ignore the different life-stage and demographic characteristics of the customers and of course those differences in terms of the profit that each customer provides to the bank.

The marketing officers of a fictional bank decided to use the available data to mine their consumer customers and divide them into groups according to their value and behavioral characteristics. They also intended to:

- **Identify the specific characteristics and needs of each segment:** The bank's objective was to identify the segment-specific characteristics and needs and to address them appropriately through:

  - Customized product offerings that would fit the requirements of each segment.
  - Service personalization according to the profile of each segment.
  - Offering the appropriate incentives to sustain and develop the relationship with the customers.

 – Developing new products and product bundles that would match the profile of each segment.

- **Analyze the competition per segment:** The marketers' intention was to study the organization's strengths and weaknesses, compared to the competition, and reveal the specific threats and opportunities in each segment.
- **Analyze each segment 's present and potential value:** Each segment does not provide the same profit to the organization. The project team intended to evaluate each segment's present and future/potential value in order to assess its importance and incorporate a more effective allocation of resources.
- **Analyze the "share of wallet" per segment:** The "share of wallet" is the proportion of the customer's total money managed by the bank. Typically, an increase of the "share of wallet" is less costly and easier than an increase of the market share by acquiring new customers. The bank's objective was to study and assess the "share of wallet" of each segment through relevant metrics (such as the cross-selling index, defined as the average number of products held) and specialized market research surveys. The findings and the insight into the needs of each segment would then be used to develop efficient strategies for an increase in the "share of wallet" of existing customers.
- **Measure the level of satisfaction and identify the most important satisfaction drivers of each segment:** Customers do not have the same expectations from their bank and their satisfaction is affected by a wide variety of factors. A customer satisfaction survey is typically conducted to measure the level of satisfaction and reveal the drivers that have an impact on satisfaction. The adopted approach included a separate satisfaction analysis for each segment to identify the particular drivers with increased importance.

Conclusively, the end goal was the creation of an effective segmentation scheme that would allow the improvement of the relationship with customers and an increase in their level of satisfaction through better coverage of their needs. Furthermore, identification of the customer typologies would also support the bank's efforts to increase its market share as well as the "share of wallet" of existing customers, through the development of targeted incentives and product offerings that could persuade more customers to transfer from the competition.

The bank initially focused on its consumer customers, excluding corporate, wholesale, and private banking customers from the particular analysis. The project involved the following steps:

1. Value-based segmentation according to marginal revenue.
2. Separation of customers into broad and clearly differentiated groups (core segments) such as "Affluent," "Mass," "Business," "Inactive," and so on,

according to business rules that summarize their general behavior. The specific definitions for assigning segments were determined after an examination of the characteristics of the entire population.

3. Further segmentation of the emerging core segments into refined sub-segments of differentiated behavior.

SEGMENTING CUSTOMERS ACCORDING TO THEIR VALUE: THE VITAL FEW CUSTOMERS

Value-based segmentation is one of the key mining tasks. It differentiates customers according to their profit contribution and can be used to prioritize customer handling. High-value customers, after being identified, should always receive top-level service in every interaction with the bank. Since they account for a significant share of the organization's profit, the bank has to build their loyalty by offering benefits and incentives that will make clear that it recognizes their significance and will discourage them from turning to competitors. These customers should be thoroughly monitored over time and any signs indicating a possible decrease in their relationship with the bank should trigger appropriate retention activities.

The marketers of the bank decided to partition their customer base according to marginal revenue, a value index calculated previously and readily available in the bank's mining data mart. At first, all customers were ordered by rank in respect of their marginal revenue and subsequently binned into groups of equal size, referred to as quantiles or tiles. Customers with zero and negative revenue contribution were distinguished and assigned to two distinct groups. Customers who cause a loss to the bank formed tile 10 and customers with no profit contribution formed tile 9. The rest of the customers were grouped into eight value segments of about 10% each.

As shown in Table 6.12, the relative revenue contribution of each derived segment does not correspond to its size. On the contrary, respective value steeply decreases from the top to the bottom segments, despite the fact that each segment has approximately the same number of customers. Thus the top 10% of the most valuable customers (tile 1) accounts for about 60% of the total bank's revenue. In a similar manner, the second most valuable segment (tile 2) also accounts for a disproportionately large part of revenue. The Pareto principle holds true in the bank's case presented here since, jointly, tiles 1 and 2 provide about 80% of the overall bank's marginal revenue. These are the vital few customers that the bank should distinguish and focus on. The bank should try to further develop the relationship with them and prevent their churn, since their possible attrition will drastically decrease the bank's profit.

Table 6.12   The derived value-based segments.

Tiles	Profitability curve			
	Number of customers (%)	Marginal revenue sum (%)	Cumulative number of customers (%)	Cumulative marginal revenue sum (%)
1	10.00	60.40	10.00	60.40
2	10.00	20.50	20.00	80.90
3	10.00	10.80	30.00	91.70
4	10.00	5.80	40.00	97.50
5	10.00	3.10	50.00	100.60
6	10.00	1.40	60.00	102.00
7	10.00	0.50	70.00	102.50
8	10.00	0.10	80.00	102.60
9	15.70	0.00	95.70	102.60
10	4.30	−2.70	100.00	99.90

The characteristics of the top-value customers also indicate the profile of the right, potentially profitable, prospects that should be the acquisition target of the bank.

The profitability curve in Figure 6.17 depicts the derived segments and their cumulative sum of marginal revenue. The large revenue differences among segments are evident. By tile 4, the line has almost reached 100% of total revenue, denoting the trivial contribution of the subsequent tiles.

USING BUSINESS RULES TO DEFINE THE CORE SEGMENTS

The customer base was initially divided into five core segments according to their general behavior. The scope was to implement a basic separation into self-evident and clearly differentiated groups, before applying more advanced clustering techniques in search of refined sub-segments within the core segments.

The final segments were defined and built by applying business rules that employed general segmentation criteria, commonly used in the banking industry, but specifically adapted to match the situation of this particular bank. Those business rules incorporated the business knowledge and experience of the bank's marketers and also reflected the findings of an initial exploratory analysis of the characteristics of the customer base. Hence, customers were partitioned in terms of their assets (six-month average balances of savings and investments), product ownership, and total number of transactions during the last six months, into the five core segments presented in Table 6.13.

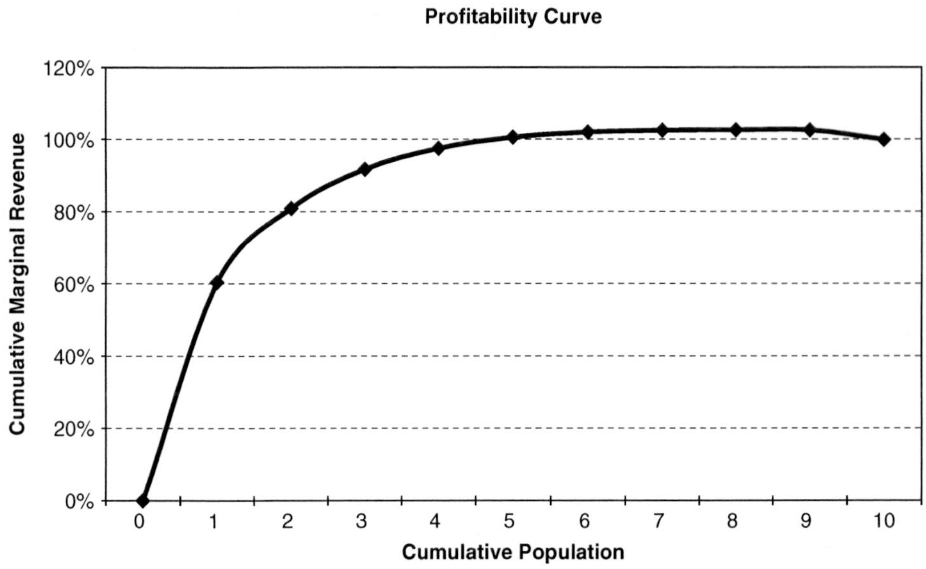

Profitability Curve

Figure 6.17   The profitability curve of the value-based segments.

Table 6.13   Bank's core segments.

Business rule	Core segment	Percentage	Segmentation criteria
1	Affluent	2	Total assets ≥$100 000
2	Affluent Mass	3	Total assets ≥$50 000 and <$100 000
3	Business Mass	11	Business product ownership
4	Pure Mass	72	Rest of customers
5	Inactive Mass	12	Maximum total balances in last six months <$50 and total number of transactions in last six months = 0

The "Affluent" segment included the top customers with the highest assets (over $100 000). Customers with assets between $50 000 and $100 000 comprised the "Affluent Mass" segment. Customers with a business loan were assigned to the "Business Mass" segment. Those customers with no transactions at all during the last six months and with low total balances (less than $50) were flagged as "Inactive" and were grouped in the "Inactive Mass" segment. Finally, the rest of the customers that did not fall into any of these categories comprised the "Pure Mass" segment.

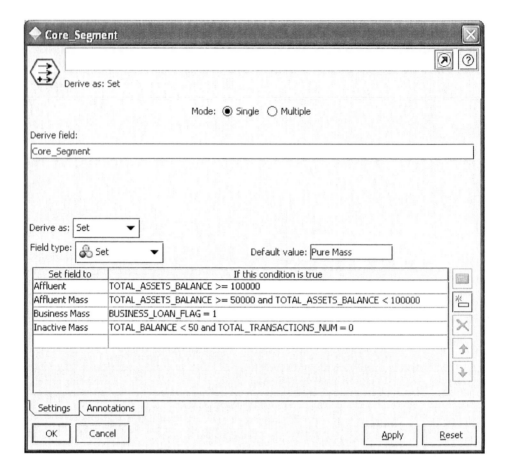

Figure 6.18 The IBM SPSS Modeler Conditional Derive node for applying the business rules and building the core segments.

The IBM SPSS Modeler Derive node, depicted in Figure 6.18, was used to build the business rules of the conditional assignment of customers to the five segments.

The "Pure Mass" segment was the largest segment containing almost three-quarters of the customer base. On the contrary, the affluent segments ("Affluent" and "Affluent Mass") jointly accounted for only 5% of total customers. The "Business Mass" and "Inactive Mass" segments contained slightly over 10% of customers each.

The "Inactive" customers merit special handling. Their dormant relationship with the bank might indicate the threat of upcoming attrition. Most probably these customers also have another bank as a main supplier of financial services; therefore, if the bank wants to retain and win them back it should initiate immediate action to stop them from turning to the competition.

The remaining segments were assigned to individual business units as a first step in building specialized marketing strategies. The efforts of the bank to gain deeper insight on its customers did not stop there. As part of a more thorough investigation, the bank decided to initially focus on the "Pure Mass" segment and to apply a clustering technique to identify the underlying distinct customer types and behaviors. The following sections present the results of this project and the behavioral sub-segments revealed for the "Pure Mass" customers.

SEGMENTATION USING BEHAVIORAL ATTRIBUTES

The aim of the sub-segmentation process was to further break down the "Pure Mass" customers into distinct smaller groups of differentiated needs and behaviors that could support the design of customized marketing efforts.

A clustering technique was applied to reveal the natural groupings of the customers with respect to their behavioral characteristics. The required information to support the training of the model was retrieved from the organization's mining data mart and MCIF. It covered six months of data and contained summarized information, aggregated at a customer level, on the following aspects of behavior:

- **Product ownership and utilization:** Each customer's balances (monthly average balances) with respect to the main product categories.
- **Types of transactions:** Relative number and volume (monthly average) per main transaction type.

These were the main segmentation dimensions. An extended list of relevant fields was prepared for the needs of the analysis, to provide a complete view of each customer in regard to the above characteristics. This list included derived attributes such as monthly averages and ratios (percentages of total balance and of total transactions) which denoted the dominant product categories and the most common types of transactions for each customer. The complete list of input fields can be found in the next section.

Selecting the Segmentation Fields

As noted above, the list of clustering inputs contained only fields related to the specific marketing objectives of the segmentation process (Table 6.14). Other informative, yet not directly relevant, fields, such as customer demographics

Table 6.14 The banking segmentation fields.

Field name	Description
Product balances:	
TOTAL_SAVINGS_BALANCE	Monthly average balance of all deposits/savings accounts
TIME_DEPOSITS_BALANCE	Monthly average balance of time deposits
TOTAL_INVESTMENTS_ BALANCE	Monthly average balance of all investment products
MORTGAGE_BALANCE	Monthly average balance of mortgage loans
TOTAL_CONSUMER_ LOANS_BALANCE	Monthly average balance of all consumer loans
CREDIT_CARDS_BALANCE	Monthly average balance of credit cards
TOTAL_ASSETS_BALANCE	Monthly average of assets (savings + time deposits + investments + insurance balances)
TOTAL_LENDINGS_ BALANCE	Monthly average of lending (consumer loans + mortgage loans + credit card balances)
PRC_TOTAL_SAVINGS_ BALANCE	Saving balances (monthly average) as a proportion of total (monthly average) balances
PRC_TIME_DEPOSITS_ BALANCE	Balances of time deposits as a proportion of total balances
PRC_TOTAL_ INVESTMENTS_BALANCE	Investment balances as a proportion of total balances
PRC_MORTGAGE_ BALANCE	Mortgage balances as a proportion of total balances
PRC_TOTAL_CONSUMER_ LOANS_BALANCE	Balances of consumer loans as a proportion of total balances
PRC_CREDIT_CARDS_ BALANCE	Credit card balances as a proportion of total balances
PRC_TOTAL_ASSETS_ BALANCE	Asset balances as a proportion of total balances
PRC_TOTAL_LENDINGS_ BALANCE	Lending balances as a proportion of total balances
Number and volume of transactions:	
DEPOSIT_NUM	Monthly average number of deposit transactions
DEPOSIT_AMOUNT	Monthly average amount of deposit transactions
WITHDRAWAL_NUM	Monthly average number of withdrawal transactions
WITHDRAWAL_AMOUNT	Monthly average amount of withdrawal transactions
CASH_ADVANCE_NUM	Monthly average number of cash advance transactions
CASH_ADVANCE_AMOUNT	Monthly average amount of cash advance transactions

Table 6.14 (*continued*)

Field name	Description
PAYMENTS_CREDIT_ CARDS_AMOUNT	Monthly average amount of credit card payments
CREDIT_CARDS_ PURCHASES_AMOUNT	Monthly average amount of credit card purchases
CREDIT_TRANSACTIONS_ NUM	Monthly average number of credit transactions (deposits, card payments, etc.)
CREDIT_TRANSACTIONS_ AMOUNT	Monthly average amount of credit transactions
DEBIT_TRANSACTIONS_ NUM	Monthly average number of debit transactions (with-drawals, card purchases, bill payments, etc.)
DEBIT_TRANSACTIONS_ AMOUNT	Monthly average amount of debit transactions
TOTAL_TRANSACTIONS_ NUM	Monthly average number of total transactions
TOTAL_TRANSACTIONS_ AMOUNT	Monthly average amount of total transactions
CREDIT_CARDS_ SPENDING_AMOUNT	Monthly average amount of credit card spending or usage (purchases + cash advances)
PRC_CREDIT_ TRANSACTIONS_NUM	Number of credit transactions as a proportion of total number of transactions
PRC_CREDIT_ TRANSACTIONS_AMOUNT	Amount of credit transactions as a proportion of total amount of transactions
PRC_DEBIT_ TRANSACTIONS_NUM	Number of debit transactions as a proportion of total number of transactions
PRC_DEBIT_TRANSACTIONS_ AMOUNT	Amount of debit transactions as a proportion of total amount of transactions
PRC_DEPOSIT_NUM	Number of deposit transactions as a proportion of total number of transactions
PRC_DEPOSIT_AMOUNT	Amount of deposit transactions as a proportion of total amount of transactions
PRC_WITHDRAWAL_NUM	Number of withdrawal transactions as a proportion of total number of transactions
PRC_WITHDRAWAL_ AMOUNT	Amount of withdrawal transactions as a proportion of total amount of transactions
PRC_CASH_ADVANCE_ NUM	Number of cash advance transactions as a proportion of total number of transactions
PRC_CASH_ADVANCE_ AMOUNT	Amount of cash advance transactions as a proportion of total amount of transactions

and information on channel utilization, were not included in the model training procedure since they would just confound the separation and lead the analytical process away from the specific business goal. However, in the end all available information of interest was taken into account during the cluster profiling and evaluation phase.

Moreover, categorical fields were also omitted from the clustering procedure, since, as mentioned in previous chapters, they tend to provide biased clustering solutions which overlook differences attributable to other inputs.

THE ANALYTICAL PROCESS

The determined segmentation process comprised two steps. At first PCA was applied to reveal the distinct data dimensions underlying the 41 inputs listed above. Then a clustering model was used to reveal the final segmentation solution.

PCA, although optional, is a useful data preparation step aimed at data reduction.

The extracted principal components, once explained and fully understood, were used as clustering inputs instead of the original fields. This was the second and final step of the analytical process: a clustering model assessed the similarities of the records/customers in terms of the revealed components and suggested the underlying customer groupings. The proposed clusters were then interpreted and evaluated, mainly in terms of their business meaning and usefulness, before concluding on the final solution adopted for the organization.

Identifying the Segmentation Dimensions with PCA/Factor Analysis

The team involved in the project selected PCA as the data reduction method. The components extracted by PCA are uncorrelated linear combinations of the original inputs. They are extracted in order of importance, with the first one carrying the largest part of the variance of the original fields. The subsequent components explain smaller portions of the total variance and are uncorrelated with each other. Moreover, the analysts involved in the project also chose to incorporate a Varimax rotation method in order to simplify interpretation of the components.

The PCA algorithm analyzed the inputs' intercorrelations and extracted 13 components which accounted for almost 85% of the variance/information of the original fields – a large step toward simplicity with a minimum loss of information. The amount of information retained by the extracted solution is summarized in Table 6.15.

This table lists the eigenvalues and the percentage of variance (plain and cumulative) explained by each extracted component. The criterion used to determine the number of components to extract was the eigenvalue (or latent root)

Table 6.15 The "variance explained" table of the PCA solution.

Component	Total variance explained		
	Initial eigenvalues		
	Total	Percentage of variance	Cumulative %
1	6.14	14.63	14.63
2	5.75	13.69	28.32
3	4.12	9.82	38.14
4	3.52	8.39	46.53
5	2.70	6.44	52.96
6	2.34	5.57	58.53
7	1.98	4.71	63.24
8	1.77	4.21	67.45
9	1.71	4.06	71.52
10	1.60	3.81	75.32
11	1.37	3.25	78.57
12	1.16	2.77	81.35
13	1.00	2.38	83.73
14	0.94	2.23	85.96
15	0.82	1.95	87.91
16	0.76	1.82	89.73
17	0.68	1.61	91.34
18	0.58	1.38	92.72
19	0.47	1.13	93.85
20	0.43	1.01	94.86
21	0.40	0.95	95.81
22	0.35	0.84	96.65
23	0.32	0.77	97.42
24	0.24	0.57	98.00
25	0.21	0.49	98.49
26	0.17	0.40	98.89
27	0.13	0.30	99.19
28	0.10	0.24	99.43
29	0.08	0.19	99.62
30	0.06	0.14	99.76
31	0.03	0.08	99.84
32	0.03	0.07	99.91
33	0.03	0.06	99.97
34	0.01	0.02	99.99
35	0.00	0.01	100.00
36	0.00	0.00	100.00
37	0.00	0.00	100.00
38	0.00	0.00	100.00
39	0.00	0.00	100.00
40	0.00	0.00	100.00
41	0.00	0.00	100.00

criterion. Only components with eigenvalues above 1 were retained and this yielded a set of 13 components, which explained 83.7% of the original information, a percentage more than satisfactory for considering the solution as representative of the initial fields.

However, the proportion of explained variance was not the only factor considered before accepting the specific solution. The extracted components are to be used as inputs to subsequent clustering models, therefore they should be understandable and clearly associated with specific behavioral aspects. After all, customers were about to be separated according to these newly derived composite measures, so it was vital that they had a crystal clear business meaning.

The component interpretation phase was based on the loadings of the 13 components. They denote the correlations of the components with the original inputs and are presented in the rotated component matrix of Table 6.16.

These loadings were examined to recognize the information conveyed by each component and facilitate its labeling. The labeled components along with a brief explanation of their meaning are presented in Table 6.17.

Segmenting the "Pure Mass" Customers with Cluster Analysis

The original fields have been temporarily left out of the segmentation procedure and substituted by the interpreted and labeled components which were used as inputs for the training of a clustering model. The clustering procedure included many trials and the application of different modeling techniques and parameter settings before concluding on the final segmentation solution to be adopted for deployment. The accepted solution was derived by a TwoStep cluster model, which, as mentioned in previous chapters, offers some useful features, such as the automatic clustering procedure that proposes the "optimal" number of clusters and the integrated handling of outliers, an option that can prevent distortion of the results due to noisy records.

The data miners involved in the project did not specify in advance the number of clusters to be created. Instead they let the algorithm propose the optimal number, between 2 and 15 clusters. The algorithm suggested a solution of four clusters and the next task was to fully understand the structure of each cluster before concluding on the usefulness of the segmentation scheme revealed.

Profiling of Segments

Standard reporting techniques were applied to reveal the data patterns of the clusters and to identify the differentiating characteristics that define them. This profiling process provided insight into the clusters and revealed the customer types and behaviors behind each grouping. Consequently, it also facilitated an

Table 6.16 The rotated component matrix with the correlations (loadings) of the components with the original PCA inputs.

Rotated component matrix

							Component						
	1	2	3	4	5	6	7	8	9	10	11	12	13
PRC CASH ADVANCE NUM	0.89												
PRC CASH ADVANCE AMOUNT	0.88												
CASH ADVANCE AMOUNT	0.87												
CASH ADVANCE NUM	0.85												
CREDIT CARDS BALANCE	0.62												
PRC DEPOSIT NUM		0.91											
PRC DEPOSIT AMOUNT		0.90											
PRC CREDIT TRANSACTIONS NUM		0.86											
PRC CREDIT TRANSACTIONS AMOUNT		0.85											
DEPOSIT NUM		0.59			0.30			0.30					
CREDIT TRANSACTIONS NUM		0.50			0.36			0.30					
PRC TOTAL ASSETS BALANCE			−0.97										
PRC TOTAL LENDINGS BALANCE			0.97										
PRC CREDIT CARDS BALANCE			0.97										
PRC TOTAL SAVINGS BALANCE			−0.97										
PRC WITHDRAWAL AMOUNT				0.83									
PRC WITHDRAWAL NUM				0.81	0.32								
PRC DEBIT TRANSACTIONS AMOUNT		−0.38		0.80									
PRC DEBIT TRANSACTIONS NUM		−0.36		0.80									
TOTAL TRANSACTIONS NUM					0.91								

(continued overleaf)

Table 6.16 (continued)

Rotated component matrix	Component												
	1	2	3	4	5	6	7	8	9	10	11	12	13
DEBIT TRANSACTIONS NUM					0.86								
WITHDRAWAL NUM				0.37	0.84								
MORTGAGE BALANCE						0.87							
TOTAL LENDINGS BALANCE						0.84							
PRC MORTGAGE BALANCE						0.83							
CREDIT CARDS PURCHASES AMOUNT							0.83						
CREDIT CARDS SPENDING AMOUNT	0.55						0.81						
PAYMENTS CREDIT CARDS AMOUNT	0.52						0.71						
DEBIT TRANSACTIONS AMOUNT								0.74					
WITHDRAWAL AMOUNT								0.72					
DEPOSIT AMOUNT								0.70					
TIME DEPOSITS BALANCE									0.85				
PRC TIME DEPOSITS BALANCE									0.81				
PRC TOTAL CONSUMER LOANS BALANCE										0.81			
TOTAL CONSUMER LOANS BALANCE										0.80			
CREDIT TRANSACTIONS AMOUNT											0.80		
TOTAL TRANSACTIONS AMOUNT											0.75		
TOTAL SAVINGS BALANCE												0.80	
TOTAL ASSETS BALANCE									0.52			0.70	
PRC TOTAL INVESTMENTS BALANCE										−0.30			0.71
TOTAL INVESTMENTS BALANCE													0.70

Table 6.17 Interpretation of the extracted principal components.

Factor	Description
1	**Cash advance**
	This component presents strong positive correlation with the fields that measure the volume (amount) and frequency (number of) cash advance transactions. Therefore it appears to be a measure of cash advance usage. Customers with high positive value on this component are expected to heavily use their credit cards for withdrawals. This component is also moderately correlated with credit card balance
2	**Deposit transactions**
	This component presents strong positive correlation with the fields that measure the frequency and volume proportion of credit transactions and deposits in particular. High positive values on this component indicate customers whose transactions are mainly deposits
3	**Lending vs. assets**
	A component that is positively correlated with the percentage of lending balances and negatively correlated with the percentage of assets balances. It denotes a contrast between assets and lending (liabilities), not necessarily in terms of balance amount but in terms of balance percentage and share in the total position. This score separates "lenders" from "depositors" and "investors." Customers with high positive component scores are expected to be "lenders" with increased liabilities. Negative component scores on the other hand signify "depositors" or "investors" with increased assets
4	**Withdrawal transactions**
	Like a mirror image of the second component, this fourth component is associated with the percentage (both frequency and volume) of debit transactions and withdrawals in particular. Therefore, withdrawals are expected to dominate the transactions of those customers who score high on this component
5	**Frequency of total transactions**
	In general, the majority of customers make more withdrawals than deposits, therefore the frequency of total transactions and withdrawals appear correlated. The fifth component measures the overall number and hence the frequency of total transactions and withdrawals in particular.
6	**Mortgage loans**
	A component associated with lending and mortgage loans in particular. It presents a strong positive correlation with mortgage balances and consequently high component scores indicate customers with increased relevant balances
7	**Credit card usage**
	The seventh component measures credit card usage, including total spending, purchases, and payments

(*continued overleaf*)

Table 6.17   (*continued*)

Factor	Description
8	**Volume of total transactions**
	The next component is associated with the volume of both deposits and withdrawals. Therefore we can conclude that it constitutes a general measure of the total transaction volume
9	**Time deposits**
	This component is concerned with time deposits. A high positive value denotes an increased time deposit balance and an increased share of these products in the customer's total position
10	**Consumer loans**
	One more component that is linked to a specific product. It is associated with consumer loan ownership. A high positive value denotes increased consumer loan balances and an increased share in the customer's total position
11	**Credit transactions volume**
	A component that measures the volume of credit transactions
12	**Savings**
	Assets and savings in particular are related to this component. Since savings balances are heavily loaded on this component, customers with increased savings balances ("depositors") are expected to score high on it
13	**Investments**
	Investment fields (balances and proportion of investment balances) present strong positive correlations with the last component. Customers with increased investment balances ("investors") should have high scores on this component

assessment of the business value and of the business opportunities offered by the solution.

The first step in the cluster profiling involved an examination of the cluster centroids and an investigation of the clusters' structure with respect to the component scores. The profiling continued with a cross-examination of the clusters with the original inputs and demographics. Specifically, the clusters' structure was examined with respect to product ownership and utilization, transactional patterns, late payments of credit cards, loan due balances (arrears), and demographics.

At first, clusters were cross-tabulated with product ownership. Table 6.18 presents the percentage of product owners within each cluster.

Cluster 1 has an increased percentage of owners of time deposits and investment products, about twice as high compared to the overall "Mass" population.

Table 6.18   Clusters and product ownership.

Clusters	Total savings (%)	Time deposits (%)	Investments (%)	Mortgage loans (%)	Consumer loans (%)	Credit cards (%)
Cluster 1	70	**12**	**19**	4	8	18
Cluster 2	**97**	1	5	1	9	26
Cluster 3	61	1	4	3	**89**	**64**
Cluster 4	**97**	**8**	9	**27**	30	40
Total	80	4	8	8	38	39

Table 6.19   Clusters and transactional patterns.

Clusters	Number of credit transactions	Credit transactions amount	Number of debit transactions	Debit transactions amount	Total credit card spending amount
Cluster 1	1	**16.828**	1	472	177
Cluster 2	1	1.552	**19**	**7.472**	240
Cluster 3	3	4.357	10	4.381	**1.001**
Cluster 4	**6**	**13.314**	5	6.514	288
Total	3	7.977	9	4.985	470

Nevertheless, their penetration in consumer loans is the lowest. On the other hand, cluster 2 is associated with savings. Almost all customers assigned to cluster 2 hold a savings account but few of them hold any other product. Cluster 4 is also characterized by savings accounts. But this cluster also presents an increased percentage of mortgage loans and time deposit owners. Finally, cluster 3 has the highest proportion of borrowers. Almost 90% of its members have a consumer loan with the bank and almost 70% are credit card holders.

The revealed clusters also show diverse transactional patterns, as given in Table 6.19.

Cluster 1 has the highest average volume (amount) of credit transactions, although the respective frequency is rather low. On the contrary, cluster 2 customers are the heaviest and most frequent debit transactions customers. As we will see in the next table, most of them have their salary deposited in the bank, thus they probably have to make regular withdrawals from their payroll savings

Table 6.20 Clusters and main customer attributes.

Clusters	Age	Other patterns			
		Assets ratio (as % of total balances)	Bucket max in previous year	Payroll flag (%)	Alternative channel usage (%)
Cluster 1	**55**	**88**	0	11	8
Cluster 2	39	75	0	**77**	28
Cluster 3	45	**10**	3	12	**51**
Cluster 4	45	48	1	11	**49**
Total	44	53	1	36	38

account for their everyday expenses. Members of cluster 3 are the biggest credit card spenders. their average credit card spending (purchases and cash advances) being almost double the overall average of all "Mass" customers. Finally, cluster 4 customers make many and large credit transactions. They have the highest average number and the second highest average amount of credit transactions.

The last profiling table, Table 6.20, examines the age distribution of the clusters and their differentiation on some main behavioral KPIs.

Cluster 1 contains the most mature part of the customer base with an average age of 55 years. The assets orientation of this cluster is underlined by one more indicator that expresses the assets as a percentage of the total balances. On average, assets account for about 88% of the total balances of each customer in this cluster, the highest ratio observed. Additionally, these customers seem to prefer the traditional channels for their transactions, namely the teller in the branch office and ATMs. Cluster 2 on the other hand contains younger customers with an average age below 40 years. Another distinctive characteristic of this cluster is its raised percentage of payroll customers. Almost 80% of them have a salary account at the bank.

Both clusters 3 and 4 are very close to the average age and also make much use of alternative channels for their transactions, including e-banking, SMS banking, and web banking. Almost half of their members have used alternative channels during the time period examined. However, the similarities end there. Cluster 4 has an average assets ratio as opposed to cluster 3, which has the smallest assets ratio and total balances dominated by liabilities (lending balances). Cluster 3, mainly composed of borrowers, also has the largest maximum bucket (maximum number of months in arrears during the last 12 months for all credit cards and loans), suggesting that some members of this cluster fail to pay their due amount on time.

The profiling procedure described above concluded with the naming of the derived clusters as follows:

- **Segment 1:** Depositors–investors
- **Segment 2:** Young active customers
- **Segment 3:** Borrowers–credit card users
- **Segment 4:** Full relationship customers.

THE MARKETING PROCESS

The revealed clusters passed the initial evaluation as they were considered meaningful and representative of the natural groupings underlying the particular bank's "Mass" customer base. Although the information used for profiling the clusters allowed for a thorough interpretation in terms of their behavior, it did not cover qualitative characteristics, such as the customer share, the satisfaction level, and the attrition risk of the customers. This information could not be extracted from the data residing in the organization's mining data mart and data warehouse. It could only be retrieved through marketing research. Therefore the marketers decided to collect and analyze relevant data to gain further insight into the revealed clusters that could help them to design and deploy more effective marketing plans.

A representative sample of each "Mass" behavioral segment was selected and included in the marketing research that aimed at collecting additional information. The respective results complemented the behavioral characteristics of the segments and the entire profiling information was finally integrated into the summarizing tables that follow.

The enrichment of the behavioral segments through marketing research is a generally recommended approach. The results can be leveraged to provide a deeper understanding of the data mining segments by investigating qualitative dimensions such as needs/intentions/attitudes/opinions/perceptions/satisfaction. This information can shed more light on the characteristics of the identified groups and also verify the data mining findings.

Setting the Business Objectives

The process of setting the business objectives and creating the marketing plan for each identified segment should take into account all its defining characteristics. Therefore all relevant findings, including the available results from marketing research surveys, should be put together and integrated into a complete profile for each segment.

The profile of the revealed segments for the bank case examined here is presented in a set of two tables. The first one summarizes the segments in terms of behavioral information retrieved from the organization's data mart. The second one summarizes the marketing research results and presents the estimated "share of wallet", the satisfaction level, and the attrition risk for each segment. Finally, an additional table of conclusions shows the specialized business and marketing strategies designed for each specific segment.

The first table, Table 6.21, outlines the behavioral profile and the value rank of the segments. Young active customers and borrowers–credit card users are the largest segments, each containing about 30% of the "Mass" customers. The young active customers segment is mainly composed of less valuable customers. These customers are payrolled by the bank and make many transactions (mostly withdrawals); however, their relationship with the bank is limited to a simple savings account, suggesting that they probably have another bank as a main financial partner. The borrowers–credit card users on the other hand are quite profitable and occupy the highest-value rank position. The full relationship segment accounts for 23% of the "Mass" customer base and it also has a top-value rank. Unlike young employees who seem to be new customers not tied to the bank, this segment includes customers with a more established relationship. They have average savings balances and many of them also have a mortgage loan from the bank. The depositors–investors segment is the smallest one with about 20% of the customers and a medium-value index. As described earlier and as implied by the assigned name, it includes customers with a rather conservative portfolio comprising deposits and low-risk investments. They prefer the manual, traditional channels over alternative ones.

The second table, Table 6.22, presents the main marketing research findings on each segment in terms of satisfaction, risk of defection, and "share of wallet." As expected, the full relationship customers have the highest share of their money with the bank in contrast to the young active customers, who seem to spread their wallet among different banks. The other two segments are average with respect to their "share of wallet." In terms of attrition risk, borrowers–credit card users appear to have the highest probability to churn and switch to competitors, maybe due to their relatively low satisfaction level. Moreover, because of the nature of their products, they probably tend to continuously assess their received benefits in a constant search of the market for better offers and terms that can entice them to move to the competition. At the other end are the depositors–investors and the young active customers. They seem quite settled and satisfied with the services offered and show no intention to churn. In the case of new payroll customers, their loyalty is easily explained by the type of relationship they have with the bank. They receive their salary there, so it is quite difficult to migrate to a competitor. The challenge and opportunity in their case is to further develop them as good

Table 6.21 The behavioral profile of the segments.

Number	Segment	Size (%)	Profiling tips	Value rank
1	**Depositors–investors**	19	• High savings balances • High volume of credit transactions for depositing money • High penetration in time deposits and investments • Low penetration in consumer loans • Above-average educational level • Above-average age • Usage of traditional channels	Medium
2	**Young active customers**	28	• Holders of savings accounts with average balances • Very high frequency and volume of withdrawals • Above-average educational level • Young employees starting their career • Most employees are payrolled by the bank	Low
3	**Borrowers–credit card users**	30	• Low savings balances • High withdrawal frequency • High penetration in consumer loans • Heavy credit card users • Increased bucket (some of them carry over their due amount and are months in arrears) • Increased usage of alternative channels (for paying their loans)	High
4	**Full relationship customers**	23	• Average savings balances • High penetration in time deposits • High penetration in mortgage loans • Above-average educational level • Users of alternative channels	High

Table 6.22 A summary of the market research findings on the segments.

Number	Segment	Customer share	Attrition risk	Satisfaction
1	**Depositors–investors**	Medium	Low	High
2	**Young active customers**	Low	Low	Medium
3	**Borrowers–credit card users**	Medium	High	Low
4	**Full relationship customers**	High	Medium	Medium

and truly profitable customers. The low attrition risk of the more conservative and traditional, elderly depositors–investors could be possibly explained by an understandable unwillingness to become actively involved and to re-evaluate their decisions. They settle with their current security and try to avoid the fuss of transferring their assets to a new bank.

Each segment has distinct characteristics which induce growth opportunities when tackled appropriately. All the aforementioned profiling information was put together and assessed by the marketers of the bank in order to develop tailored business objectives for each "Mass" segment. These goals are summarized in the conclusions table, Table 6.23.

The main objective for the depositors–investors was to attract an even higher share of their assets. Another goal of equal importance was to motivate them, through appropriate incentives, to use their premium credit cards more. These people should feel secure and that they are co-operating with a reliable financial organization to which they can entrust their wealth. The bank should offer them the feeling that it is guarding and protecting their assets. It can also provide investment guidelines and suggestions for low-risk investment products. These suggestions can be differentiated and tailored to their investment profile. Another objective concerning those customers was their gradual transfer from the manual, traditional channels (teller in the branch office) to less expensive, alternative ones (ATMs, phone banking, etc.).

A large part of the young active customers segment has a rather new and limited one-product relationship with the bank. Sustaining and further developing this relationship was set as the first priority by offering loyalty promoting products. This is a diverse and dynamic segment that can serve as the pool for new full relationship customers. They are probably heavy consumers. Therefore, they can be motivated to increase their credit card usage. Some of them may be facing even bigger needs and challenges due to a possible change in their life, such as marriage or a newborn baby. Consequently, some of them may also turn out to be

Table 6.23 Using the segments to set distinct business objectives.

Number	Segment	Objectives	Focus	Brand image communicated
1	**Depositors–investors**	• Additional savings with the competition • Premium credit cards • Low-risk investment products • Introducing them to alternative, non-manual channels	Focus on people and differentiated treatment	A credible financial organization that can guard and expand customer assets
2	**Young active customers**	• Credit card usage incentives • Consumer loans • Mortgage • Internet banking • Automatic payments	Focus on transactional convenience and development of their relationship with the bank	A modern bank with a range of innovative products that offers transactional convenience and is developing with its customers
3	**Borrowers–credit card users**	• Loyalty incentives • Specialized credit cards • Consumer loan for credit card holders with debts • Mortgage loans (considering credit score) • Internet banking • Automatic payments	Focus on product and pricing	A bank that understands the needs of its customers and helps them reach their goals by offering a variety of specialized lending products at competitive rates and with usage incentives
4	**Full relationship customers**	• Loyalty incentives and rewards • Mortgage loans (for the non-owners) • Internet banking • Automatic payments	Focus on relationship pricing and reward	An established bank that offers a wide range of competitive products and rewards its loyal customers

good prospects for lending products such as a consumer loan or even a mortgage. These customers are heavily involved with transactions. Therefore they should be encouraged to make use of the bank's alternative channels. They should be aware of all the options they have for taking care of their business without visiting a branch. The self-service options (online banking, automatic payments, etc.) should be communicated to them and their transactional convenience should be regarded as a main priority.

Borrowers–credit card users are high-value customers with a high attrition risk and low satisfaction rates. Their handling included the offering of loyalty incentives to prevent possible migration to competitors. These customers are price sensitive and keep on gathering information concerning consumer lending products on the market. Thus, the bank should focus on their product bundle and pricing to reduce their attrition risk and increase their satisfaction level. The bank should further analyze their needs and behaviors and offer them specialized products that better match their wants and requirements. For instance, the mining of credit card usage data can identify the distinct purchase behaviors of credit card holders which can then be addressed by specialized cards and reward schemes. Likewise, a personal loan solution can be proposed to credit card holders who have problems with the payment of their dues. Many of these people are already familiar with the alternative channels. To facilitate their interactions with the bank and their payments, they should be introduced to the benefits of Internet banking and automatic payments. This segment can prove to be a good target group for promoting a mortgage loan. Considering that some of them are delinquent debtors, their credit score should be taken into account before making such an offer.

Full relationship customers are the backbone of the organization. They are high-value customers who have a strongly established relationship with the bank. These people should feel that they are specially treated and that the bank fully understands their needs and handles their possible complaints in a timely manner. They should perceive their importance in every interaction with the bank. Their loyalty should be rewarded with incentives, special offers, tailored product bundles, and special pricing as a sign of recognition of their significance and as an attempt to further develop their relationship with the bank. Alternative transaction channels could be further promoted to these customers. Since the bank is the main financial partner for most of them, their confidence and trust may consist of a cross-selling opportunity for a mortgage loan.

SEGMENTATION IN RETAIL BANKING: A SUMMARY

The marketers of a bank decided to segment their consumer customer base in order to better understand their customers and more effectively tackle their

Table 6.24   An outline of the bank's segmentation scheme.

Rule	Segment	Percentage	Definition
1	Affluent	2	Total assets $\geq \$100\,000$
2	Affluent Mass	3	Total assets $\geq \$50\,000$ and $< \$100\,000$
3	Business Mass	11	Business loans or other business services customers
4	Pure Mass	72	(a) Depositors–investors (14%) (b) Young active customers (20%) (c) Borrowers–credit card users (22%) (d) Full relationship customers (16%)
5	Inactive Mass	12	Max total balances in last six months $<\$50$ and total number of transactions in last six months $= 0$

differentiated needs. Their approach combined business rules and automatic clustering algorithms. The main dimensions used for grouping the customers were their total assets, product utilization (balances), and transactional patterns. The profiling of the segments was supplemented with marketing research findings and demographic information. The revealed segments are listed in Table 6.24.

Customers with a minimum of $50 000 of assets comprised the "Affluent" and "Affluent Mass" segments. These segments jointly accounted for 5% of the customer base. Customers with business products such as business loans, business accounts, and business services were considered as a special case and assigned to a distinct "Business Mass" segment.

The "Pure Mass" segment included the rest of the clients. It was the bank's most crowded segment with over 70% of the total customer base. Due to its diversity and the different customer types that it was believed to encompass, the marketers of the organization decided to further segregate this segment according to behavioral criteria. As a result, four behavioral sub-segments were identified, separating "Pure Mass" customers into depositors–investors, young active customers, borrowers–credit card users, and full relationship customers.

"Inactive" customers, with low asset balances and transaction dormancy, were distinguished from the rest and made up a segment of their own. These customers were set apart to investigate the reasons for inactivity and for inclusion in upcoming reactivation campaigns.

The primary goal of this project was to gain insight on the customers and use this knowledge to design differentiated business strategies for better customer handling. The first move in this direction by the bank's executives was the assignment of the management of the derived segments to distinct managers and dedicated teams. The tasks of segmentation managers included the monitoring of the evolution of each segment through derived KPIs and the design of specialized marketing strategies.

CHAPTER SEVEN

Segmentation Applications in Telecommunications

MOBILE TELEPHONY

It is a jungle out there for mobile telephony network operators, with an environment of strong competition, especially in the case of mature markets. Offering high-level quality services is essential for becoming established in the market. In times of rapid change and fierce competition, focusing only on customer acquisition, which is nevertheless becoming more and more difficult, is not enough. Inevitably, organizations have to also work on customer retention and on gaining a larger "share of customers" instead of trying only to gain a bigger slice of the market. Growth from within is sometimes easier to achieve and equally as important as winning customers from competitors.

Hence, keeping customers satisfied and profitable is a one-way street to success. In order to achieve this, operators have to focus on customers and understand their needs, behaviors, and preferences. Behavioral segmentation can help in the identification of different customer typologies and in the development of targeted marketing strategies.

Nowadays, customers can choose from a huge variety of services. The days of voice-only calls are long gone. Mobile phones are communication centers and it is up to users to select the way of usage that suits their needs. People can communicate via SMS and MMS messages. They can use their phones to connect to the Internet, to send e-mails, to download games or ringtones, and to communicate with friends and family when they travel abroad. Mobile phones are perceived differently by various people. Some customers use them only in rare circumstances and mainly

Data Mining Techniques in CRM: Inside Customer Segmentation K. Tsiptsis and A. Chorianopoulos
© 2009 John Wiley & Sons, Ltd

for receiving incoming calls. Others are addicted to their devices and cannot live without them. Some treat them as electronic gadgets, while for others they are a tool for their work.

Clearly this multitude of potential choices results in different usage patterns and typologies. Once again the good news is that usage is recorded in detail: CDRs (Call Detail Records) are stored, providing a detailed record of usage. They contain information on all types of calls. When aggregated and appropriately processed, they can provide valuable information for behavioral analysis.

All usage history should be stored in the organization's mining data mart and MCIF, as described in the respective chapter of this book. Information on frequency and intensity of usage for each type of call (voice, SMS, MMS, Internet connection, etc.) should be taken into account when trying to identify the different patterns of behavior. In addition, information such as the day/time of calls (work days versus non-work days, peak versus off-peak hours, etc.), roaming usage, direction of calls (incoming versus outgoing), and origination/destination network type (on-net, off-net, etc.) could also contribute to the formation of a rich segmentation solution.

In this section we present a segmentation example from the mobile telephony market. The marketers of a mobile telephony network operator decided to segment their customers according to their behavior. They used all the available usage data to reveal the natural groupings in their customer base. Their goal was to fully understand their customers in order to:

- Develop tailored sales and marketing strategies for each segment.
- Identify distinct customer needs and behaviors and proceed to the development of new products and services, targeting the diverse usage profiles of their customers. This could directly lead to increased usage on behalf of existing customers but might also attract new customers from the competition.

Moreover, the operator's marketers also decided to segment their customers according to revenue and to distinguish high-value from medium- and low-value customers. Their intention was to use this information to incorporate prioritization strategies and handle each customer accordingly.

MOBILE TELEPHONY CORE SEGMENTS – SELECTING THE SEGMENTATION POPULATION

Mobile telephony customers are typically categorized in core segments according to their rate plans and the type of relationship with the operator. The first segmentation level differentiates residential (consumer) from business customers. Residential customers are further divided into two, postpaid and prepaid:

- **Postpaid – contractual customers:** Customers with mobile phone contracts. Usually these customers comprise the majority of the customer base. They have a contract and a long-term billing arrangement with the network operator for the services received. They are billed on a monthly basis and according to the traffic of the past month, hence they have unlimited credit.
- **Prepaid customers:** These customers do not have a contract-based relationship with the operator and buy credit in advance. They do not have ongoing billing and they pay for the services before actually using them.

Additionally, business customers are further differentiated according to their size into Large Business – Corporate, SME (Small and Medium Enterprise), and SOHO (Small Office, Home Office) customers.

The typical core segments in mobile telephony are depicted in Figure 7.1.

The objective of the marketers was to enrich the core segmentation scheme with refined sub-segments. Therefore they decided to focus their initial segmentation attempts exclusively on residential postpaid customers. Prepaid customers need a special approach in which attributes like the intensity and frequency of top-ups (recharging of their credits) should be taken into account. Business customers also need a different handling since they comprise a completely different market. It is much safer to analyze these customers separately, with segmentation approaches like the ones (value based, size based, industry based, etc.) presented in Chapter 5.

Moreover, only MSISDNs (telephone numbers) with current status active or in suspension (due to payment delays) were included in the analysis. Churned (voluntary and involuntary churners) MSISDNs have been excluded from the

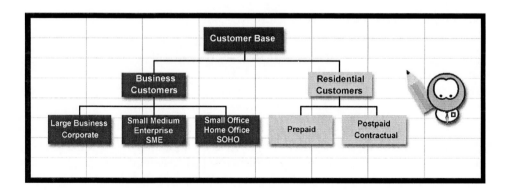

Figure 7.1   Core segments in mobile telephony.

analysis. Their "contribution" is crucial in the building of a churn model but they have nothing to offer in a segmentation scheme mainly involving phone usage. In addition, the segmentation population was narrowed down even more by excluding users with no incoming or outgoing usage within the past three months. These users have been flagged as inactive and selected for further examination and profiling. They could also form a target list for an upcoming reactivation campaign, but they do not have much to contribute to a behavioral analysis since, unfortunately, inactivity is their only behavior at the moment.

BEHAVIORAL AND VALUE-BASED SEGMENTATION – SETTING UP THE PROJECT

The methodological approach followed was analogous to the general framework presented in detail in the relevant chapter. In this section we just present some crucial points concerning the project's implementation plan, which obviously affected the whole application.

Customers may own more than one MSISDN, which may be used in a different manner to cover different needs. In order to capture all the potentially different usage behaviors of each customer, it was decided to implement the behavioral segmentation at MSISDN level. Therefore, relevant input data have been aggregated accordingly and the derived cluster model assigned each MSISDN to a distinct behavioral segment.

A two-way approach was decided for the value-based segmentation. The value segments were identified at both an MSISDN and a customer level, providing a complete view of profitability. Since the methodological approach does not depend on the level of the analysis, only the MSISDN value segmentation will be described here.

The behavioral segmentation implementation included the application of a data reduction technique (PCA in particular) to reveal the distinct dimensions of information, followed by a clustering technique to identify the segments. Once again the mining data mart tables, presented in detail in the corresponding chapter, comprised the main sources of input data. Table 7.1 outlines the main usage aspects that were covered.

On the other hand, value-based segmentation relies only on a single field. It does not need the application of a data mining algorithm either. It only involves a simple sorting of records (MSISDNs or customers) according to a profitability index and an assignment to corresponding groups. In the case presented here, it is assumed that the respective value index is already calculated and stored in the organization's MCIF. The marketers of the telephone operator used the already calculated marginal average revenue per user (MARPU), a marginal profitability

Table 7.1   Mobile telephony usage aspects that could be examined
in behavioral segmentation.

Number of calls	*Information by*	*Core service type*	• Voice • SMS • MMS • Internet • WAP • …
Minutes/traffic		*Call direction*	• Incoming • Outgoing
		Type of origination/ destination network	• On-net • Competitor A, B, C … • Fixed • International • Roaming
Community		Call hour	• Peak • Off-peak
		Call day	• Work • Non-work

index that denotes the revenue minus the costs of products/services, for building the value segments.

Both implementations included a phase of extensive profiling of the revealed segments. All available information was used in this phase, including demographic data and contract information details.

The analysis was based on data from the last six months, a time frame that generally ensures the capturing of stable, non-volatile behavioral patterns instead of random or outdated ones. Summarizing fields (sums, counts, percentages, averages, etc.) covering the six-month period under study were used as clustering fields for the behavioral segmentation. For the same reasons, the MARPU six-month average was used for the value-based segmentation, instead of one month's profitability data.

Table 7.2 summarizes the setup for the two segmentation implementations. The list of input fields is presented in the next section.

SEGMENTATION FIELDS

The list of candidate inputs for behavioral segmentation initially included all usage fields contained in the mining data mart and the MCIF. In a later stage

Table 7.2 Mobile telephony segmentation: summary of implementation methodology.

Critical point	Decision
Segmentation level	• Behavioral segmentation: MSISDNs – subsequent aggregation of results at a customer level • VBS: two-way approach, MSISDN and customer level
Segmentation population	• Residential postpaid MSISDNs, currently active (or suspended) with usage activity during the last three months
Data sources	• Mining data mart and MCIF
Time frame of used data	• Six months
Segmentation fields	• Behavioral segmentation: usage KPIs contained in the MCIF (percentages, averages, etc.) summarizing usage of the last six months • VBS: value index (MARPU) summarizing MSISDN's profitability
Profiling fields	• All fields of interest: usage KPIs for the last six months, demographics, contract details, etc.
Technique used	• Behavioral segmentation: clustering technique to reveal segments • VBS: binning

and according to the organization's specific segmentation priorities, the marketers arrived at a limited list of clustering fields, which is presented in Table 7.3. Obviously this list cannot be considered a "magic" approach that covers all the needs of all organizations. It represents the approach adopted in the specific implementation, but it also outlines a general framework of potential types of fields that could prove useful in similar applications.

The selected fields indicate the marketers' orientation toward a segmentation scheme that would capture usage differences in terms of preferred *type of calls* (voice, SMS, MMS, Internet, etc.), *roaming* usage (calls made in a foreign country), frequency of *international calls* (calls made in home country to international numbers), call *day* (peak vs. off-peak), and *hour* (work vs. non-work day) of usage.

For the needs of value-based segmentation, the only field used was the value index. The mobile telephony operator had already implemented a relevant project for the calculation of ARPU (Average Revenue Per User) and MARPU for each MSISDN. ARPU is a revenue-only index. It does not take into account costs, but covers all types of revenues (e.g., monthly fees, activation fees, outgoing usage revenue, incoming usage revenue, i.e., interconnection costs paid by other operators to recipient's operator) and possible discounts. In MARPU, however, all costs of providing products and services (e.g., interconnection costs, roaming costs, acquisition/renewal costs such as handset costs, distribution/dealers' costs, retention costs) are taken into account and subtracted from ARPU.

The full list of used fields is presented in Table 7.3.

Table 7.3 Mobile telephony segmentation fields.

Field name	Description
Fields for value-based segmentation:	
MARPU	Monthly average of MARPU value index
Clustering fields for behavioral segmentation:	
Community:	
OUT_COMMUNITY_TOTAL	Total outgoing community: monthly average of distinct phone numbers that the holder called (includes all call types)
OUT_COMMUNITY_VOICE	Outgoing voice community
OUT_COMMUNITY_SMS	Outgoing SMS community
PRC_OUT_COMMUNITY_VOICE	Percentage of outgoing voice community: outgoing voice community as a percentage of total outgoing community
PRC_OUT_COMMUNITY_SMS	Percentage of outgoing SMS community
IN_COMMUNITY_VOICE	Incoming voice community
IN_COMMUNITY_SMS	Incoming SMS community
IN_COMMUNITY_TOTAL	Total incoming community
PRC_IN_COMMUNITY_VOICE	Percentage of incoming voice community
PRC_IN_COMMUNITY_SMS	Percentage of incoming SMS community
Number of calls by call type:	
VOICE_OUT_CALLS	Monthly average of outgoing voice calls
VOICE_IN_CALLS	Monthly average of incoming voice calls
SMS_OUT_CALLS	Monthly average of outgoing SMS calls
SMS_IN_CALLS	Monthly average of incoming SMS calls
MMS_OUT_CALLS	Monthly average of outgoing MMS calls
EVENTS_CALLS	Monthly average of event calls
INTERNET_CALLS	Monthly average of Internet calls
TOTAL_OUT_CALLS	Monthly average of outgoing calls (includes all call types)
TOTAL_IN_CALLS	Monthly average of incoming calls (includes all call types)
PRC_VOICE_OUT_CALLS	Percentage of outgoing voice calls: outgoing voice calls as a percentage of total outgoing calls
PRC_SMS_OUT_CALLS	Percentage of SMS calls
PRC_MMS_OUT_CALLS	Percentage of MMS calls
PRC_EVENTS_CALLS	Percentage of event calls
PRC_INTERNET_CALLS	Percentage of Internet calls

(continued overleaf)

Table 7.3 (*continued*)

Field name	Description
Minutes/traffic by call type:	
VOICE_OUT_MINS	Monthly average number of minutes of outgoing voice calls
VOICE_IN_MINS	Monthly average number of minutes of incoming voice calls
EVENTS_TRAFFIC	Monthly average of events traffic
GPRS_TRAFFIC	Monthly average of GPRS traffic
International calls/roaming usage:	
OUT_CALLS_ROAMING	Monthly average of outgoing roaming calls (calls made in a foreign country)
OUT_MINS_ROAMING	Monthly average number of minutes of outgoing voice roaming calls
PRC_OUT_CALLS_ROAMING	Percentage of outgoing roaming calls: roaming calls as a percentage of total outgoing calls
OUT_CALLS_INTERNATIONAL	Monthly average of outgoing calls to international numbers (calls made in home country to international numbers)
OUT_MINS_INTERNATIONAL	Monthly average number of minutes of outgoing voice calls to international numbers
PRC_OUT_CALLS_ INTERNATIONAL	Percentage of outgoing international calls: outgoing international calls as a percentage of total outgoing calls
Usage by day/hour:	
OUT_CALLS_PEAK	Monthly average of outgoing calls in peak hours
OUT_CALLS_OFFPEAK	Monthly average of outgoing calls in off-peak hours
OUT_CALLS_WORK	Monthly average of outgoing calls in work days
OUT_CALLS_NONWORK	Monthly average of outgoing calls in non-work days
PRC_OUT_CALLS_PEAK	Percentage of outgoing calls in peak hours
PRC_OUT_CALLS_OFFPEAK	Percentage of outgoing calls in non-peak hours
PRC_OUT_CALLS_WORK	Percentage of outgoing calls in work days
PRC_OUT_CALLS_NONWORK	Percentage of outgoing calls in non-work days
IN_CALLS_PEAK	Monthly average of incoming calls in peak hours
IN_CALLS_OFFPEAK	Monthly average of incoming calls in off-peak hours
IN_CALLS_WORK	Monthly average of incoming calls in work days
IN_CALLS_NONWORK	Monthly average of incoming calls in non-work days

Table 7.3   (*continued*)

Field name	Description
PRC_IN_CALLS_PEAK	Percentage of incoming calls in peak hours
PRC_IN_CALLS_OFFPEAK	Percentage of incoming calls in non-peak hours
PRC_IN_CALLS_WORK	Percentage of incoming calls in work days
PRC_IN_CALLS_NONWORK	Percentage of incoming calls in non-work days
Days with usage:	
DAYS_OUT	Monthly average number of days with any outgoing usage
DAYS_IN	Monthly average number of days with any incoming usage
Average call duration:	
ACD_OUT	Average duration of outgoing voice calls (in minutes)
ACD_IN	Average duration of incoming voice calls (in minutes)
Other usage fields – profiling fields	
Contract information fields – profiling fields (for instance, tenure, rate plan, acquisition channel, payment method, handset category, etc.)	
Customer information fields – profiling fields (customer demographics)	

Data Audit

Before beginning any data mining project it is necessary to perform a health check on the data to be mined. Initial data exploration may involve looking for missing data and checking for inconsistencies, identifying outliers, and examining the field distributions with basic descriptive statistics and charts like bar charts and histograms. IBM SPSS Modeler offers a tool called Data Audit (Figure 7.2) that performs all these preliminary explorations and allows users to understand the data and spot potential abnormalities.

As clustering algorithms are very sensitive to extreme values, we should thoroughly examine the validity of the input data before beginning the model training. A common pitfall, for instance, is the inclusion of irrelevant populations, like members of staff or business customers, in the residential customer base, when the objective is to segment consumer customers. These misplaced records may behave exceptionally different from the population of interest, resulting in outlier values, which may mislead the analysis. Instead of discovering the mistakes too late, it is always preferable to perform exhaustive preliminary checks on the data in advance.

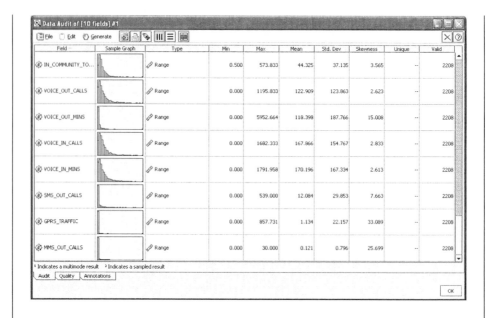

Figure 7.2   The Data Audit tool in IBM SPSS Modeler for initial data exploration.

VALUE-BASED SEGMENTATION

All customers are not equal. Some are more valuable than others. Identifying those valuable customers and understanding their importance should be considered a top priority for any organization.

Value-based segmentation is one of the key tools for developing customer prioritization strategies. It can enable service-level differentiation and prioritization of churn prevention activities. The marketers at the mobile telephony network operator decided to develop a value-based segmentation scheme in order to assign each customer to a segment according to his or her value. They believe that this segmentation will enable them to separate valuable customers from the rest and provide insight into their differentiating characteristics.

The value-based segmentation was based on the previously calculated marginal revenue for each MSISDN. Thus, all MSISDNs were ranked according to their MARPU. The ordered records were divided (binned) into equal-sized subsets, or quantiles, and these tiles were then used to construct the value segments.

What If a Value Index Is Not Available?

If there is no calculated valid value index, usage fields, known by business users to be highly related to value, could be used as temporary substitutes for the value measure, so that value segmentation could proceed. For instance, the field of monthly average number of call minutes in telecommunications may not be ideal component for value segmentation, but it could be used as a workaround.

In value-based segmentation the number of binning tiles to be constructed depends on the specific needs of the organization. The derived segments are usually of the form highest $n\%$, medium high $n\%$, medium low $n\%$, low $n\%$. In general, it is recommended to select a sufficiently rich and detailed segmentation level, especially in the top groups, in order to discern the most valuable customers. A detailed segregation level may not be required for the bottom of the value pyramid.

The segmentation bands selected by the marketers were the following:

1. **Platinum:** The top 1% MSISDNs with the highest MARPU
2. **Gold:** The top 4% MSISDNs with the second highest MARPU
3. **Silver:** 15% of MSISDNs with high MARPU
4. **Bronze:** 40% of MSISDNs with medium MARPU
5. **Mass:** 40% of MSISDNs with the lowest MARPU.

Since customers with no usage activity have been excluded from the procedure, a sixth segment, not listed but implied, is the one composed of "Inactive" customers.

The whole value-based segmentation procedure is graphically depicted in Figure 7.3.

IBM SPSS Modeler offers a very useful tool (node) called Binning that can automate the above procedure. It performs ranking and creates the quantiles in one step. The respective menu is shown in Figure 7.4.

In order to implement the value-based segmentation, the marketers selected MARPU as the "Bin field," "Tiles (equal count)" as the binning method, and "Percentile (100)" for the initial grouping of the MSISDNs. These options created bands of 1% according to the ranked MARPU.

The derived bands were then been processed with a "Reclassify" IBM SPSS Modeler node, in order to refine the solution and regroup the bands into the desired segments, as indicated in Figure 7.5.

The "MARPU_TILE100" field is the one created by the "Binning" node which denotes the assignment into 100 bands of 1%. These bands were recoded

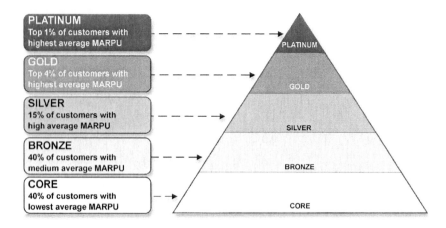

Figure 7.3   An illustration of the value-based segmentation procedure.

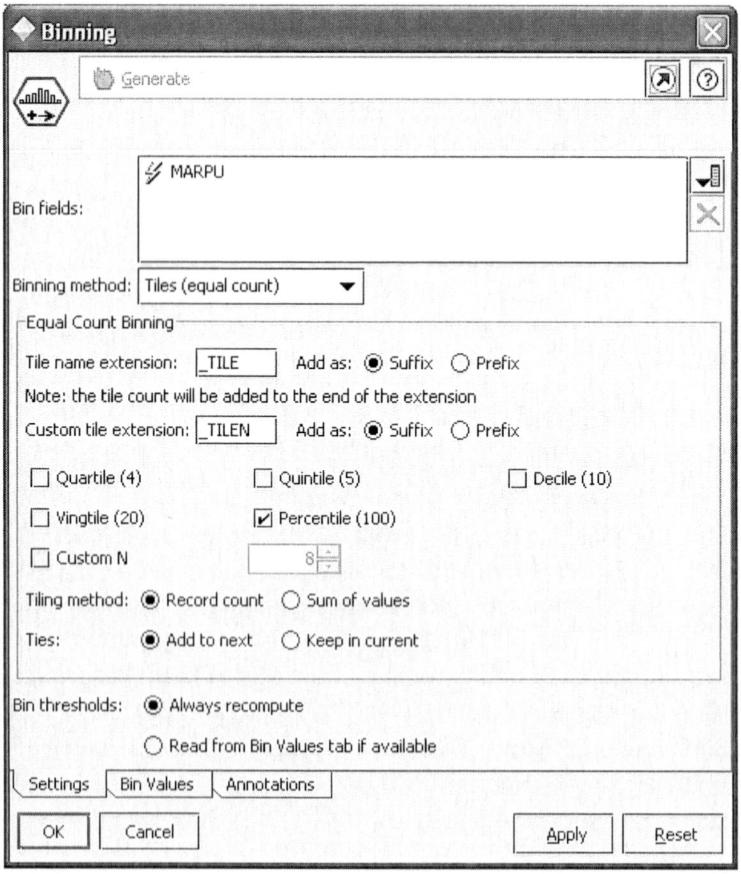

Figure 7.4   The IBM SPSS Modeler Binning Node for value-based segmentation.

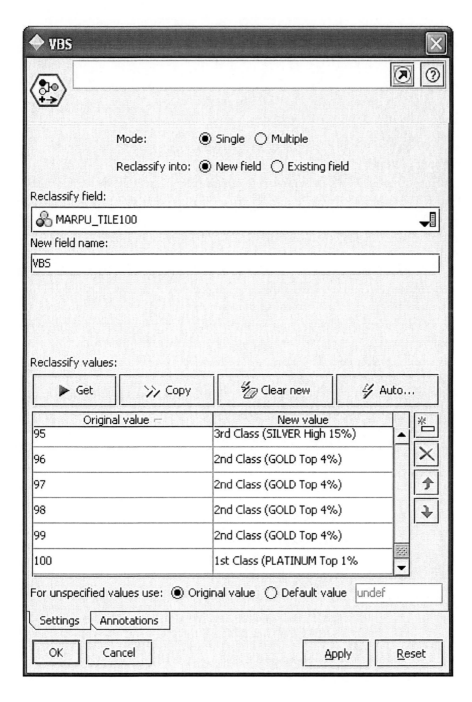

Figure 7.5   The IBM SPSS Modeler Reclassify node used for refining the value quantiles.

into broader groups through the "Reclassify" node which mapped the original quantiles into the value-based segments. The derived "VBS" field is the final result and indicates the value-based assignment for each user.

VALUE-BASED SEGMENTS: EXPLORATION AND MARKETING USAGE

The next step before making use of the derived segments was the investigation of their main characteristics.

The initial goal of this segmentation was to capture the assumed large-scale differences between customers, in terms of value. Thus, the marketers began to investigate this hypothesis by examining what each segment represents in terms of revenue. Table 7.4 presents the number of MSISDNs and the percentage of total MARPU for each value segment.

This table shows that a substantial percentage of revenue comes from a disproportionately small number of high-value users. Almost half of the operator's revenue arises from the top three value segments. Thus, 20% of the most valuable users account for about 50% of the total MARPU. On the other hand, low-value users which comprise the mass (bottom 40%) segment provide a disappointing 13% of the total revenue. The comparison of mean MARPU values also underlines the large-scale differences among segments. The mean MARPU value in the top 1% segment is more than 17 times higher than that for the bottom 40%. These findings, although impressive, are far from reality. On the contrary, in other markets, in banking for example, things are even more polarized, with an even larger part of the revenue originating from the top segments of the value pyramid.

Table 7.4   Value-based segments and MARPU.

		Percentage of MSISDNs	Sum % of MARPU	Mean MARPU
Value-based segments	First class (PLATINUM Top 1%)	1	6	219
	Second class (GOLD Top 4%)	4	14	129
	Third class (SILVER High 15%)	15	30	74
	Fourth class (BRONZE Medium 40%)	40	37	34
	Fifth class (MASS Low 40%)	40	13	12
Total		100	100	37

The Pareto Principle

Quite often, when developing a value-based segmentation scheme, we will come across the Pareto principle. This principle, also known as the 80–20 rule, states that in many cases 80% of the effects come from 20% of the causes (Figure 7.6). In business, the principle simply means that some

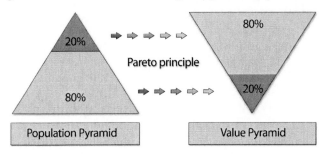

Figure 7.6 The Pareto principle and value-based segmentation.

customers are more valuable and more important than others. Keeping this part of the customer base satisfied and loyal is vital for the company's sustainability and growth.

These vast differences in revenues, as expected, are also reflected in the intensity of usage of the different services. Table 7.5 summarizes some of the main usage KPIs by segment.

Table 7.5 Value-based segments and usage of services.

		Total out calls	Voice out calls	SMS out calls	Voice out minutes	Days out
Value-based segments	First class (PLATINUM Top 1%)	719	715	81	645	28
	Second class (GOLD Top 4%)	447	426	72	423	29
	Third class (SILVER High 15%)	265	251	33	233	28
	Fourth class (BRONZE Medium 40%)	136	123	19	103	25
	Fifth class (MASS Low 40%)	69	65	7	55	18
Total		147	137	19	122	23

Once again, the interpretation is quite clear since the contrast in usage between the top and bottom segments is very intense. In almost all of these KPIs (for instance, total outgoing calls, total outgoing voice calls and minutes) users at the top of the value pyramid present values up to 10 times higher than those at the bottom. For example, the average number of total outgoing calls reaches 700 calls per month for top users, while this value does not exceed 70 calls per month among mass users.

The two-fold value segmentation project concluded by developing an analogous segmentation scheme at a customer level, at the end of which each customer was allocated to a corresponding value group.

The business benefits from the identification of value segments are prominent. The implemented segmentation can provide valuable help to the marketers in setting appropriate objectives for their marketing actions according to customer value. High-value customers are the heart of the organization. Their importance should be recognized and rewarded. They should perceive their importance every time they interact with the operator. Preventing the defection of such customers is vital for any organization. Identifying valuable customers at risk of attrition should trigger an enormous effort to avoid losing them to competitors. For medium- and especially low-value customers, marketing strategies should focus on driving up revenue through targeted cross- and up-selling campaigns to make it approach that of high-value customers.

Use of Value Segments in Acquisition Campaigns

Value-based segments can provide useful information for the development of effective acquisition models. Acquisition campaigns aim at increasing market share through expansion of the customer base to include customers new to the market or drawn from competitors. In mature markets there is fierce competition for acquiring new customers. Each organization incorporates aggressive strategies, massive advertisements, and attractive discounts.

Predictive models can be used to guide customer acquisition efforts. However, a typical difficulty with acquisition models is the availability of input data. The amount of information available on people who do not yet have a relationship with the organization is generally limited compared to information on existing customers. Without data one cannot build predictive models. Thus data on prospects must be collected. Most often, buying data on prospects at an individual or post code level can resolve this issue.

The usual approach in such cases is to run a test campaign on a random sample of prospects, record their responses, and analyze them with predictive models (classification models like decision trees, for example) in order to identify the profiles associated with increased probability of accepting an offer. The derived models can then be used to score all prospects in terms of acquisition probability. The tricky part in this method is that it requires the rollout of a test campaign to record the prospects' responses in order to train the respective models.

However, an organization should not chase just any customer, but should focus instead on new customers with value prospects. Therefore, a classification model can be built to identify the profile of the existing high-value customers and extrapolate it to the list of prospects to discern those with similar characteristics. The key to this process is to build a model using only input data that are also available in external lists of contacts. For example, if only demographic information is available for prospects, the respective model should be trained only with such data.

The latter approach is illustrated in Figure 7.7.

Prospects with high acquisition propensities and increased profit possibilities should be the first to receive an acquisition offer.

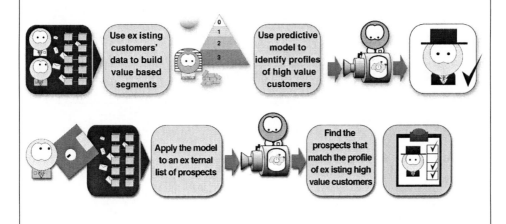

Figure 7.7   Use of value-based segments in acquisition modeling.

PREPARING DATA FOR CLUSTERING – COMBINING FIELDS INTO DATA COMPONENTS

The number of original segmentation fields exceeds 50. Using all these fields as direct input in a clustering algorithm will produce a complicated solution. Therefore the approach followed was to incorporate a data reduction technique to reveal the underlying main data dimensions, prior to clustering.

This approach was adopted to eliminate the risk of deriving a biased solution due to correlated original inputs. Moreover, it also ensures a balanced solution, to which all data dimensions contribute equally, and it simplifies the tedious procedure of understanding the clusters.

Specifically, a PCA model with Varimax rotation was applied to the original segmentation fields. The solution finally selected after many trials included 12 extracted components and was based on the "Eigenvalues over 1" criterion. The main reason for retaining this solution was the fact that it produced a relatively low number of meaningful components without sacrificing much of the information in the initial fields.

Before using the derived components and substituting more than 50 fields for just a dozen ones, the data miners of the organization wanted to be sure that the PCA solution carried over most of the original information. Therefore, they started to examine the model results by looking at the table of "variance explained," Table 7.6.

Table 7.6   Deciding the number of extracted components by examining the variance explained.

	Total variance explained		
Components	**Eigenvalue**	**Percentage of variance**	**Cumulative %**
1	**14.85**	**27.50**	**27.50**
2	**7.59**	**14.06**	**41.55**
3	**4.36**	**8.07**	**49.63**
4	**3.94**	**7.30**	**56.92**
5	**2.27**	**4.21**	**61.13**
6	**1.87**	**3.45**	**64.58**
7	**1.82**	**3.37**	**67.96**
8	**1.71**	**3.18**	**71.14**
9	**1.56**	**2.89**	**74.03**
10	**1.42**	**2.63**	**76.66**
11	**1.34**	**2.48**	**79.14**
12	**1.10**	**2.04**	**81.17**
13	0.98	1.81	82.98
14	0.93	1.72	84.73
15	0.86	1.60	86.33
16	0.77	1.42	87.75
17	0.72	1.33	89.08
⋮	⋮	⋮	⋮
54	0.00	0.00	100.00

Table 7.7 Understanding and labeling the components through the rotated component matrix.

| | Rotated component matrix | | | | | | | | | | | |
| | Components | | | | | | | | | | | |
	1	2	3	4	5	6	7	8	9	10	11	12
VOICE IN CALLS	0.93											
TOTAL IN CALLS	0.93											
IN CALLS PEAK	0.91											
IN COMMUNITY VOICE	0.88											
IN COMMUNITY TOTAL	0.88											
IN CALLS WORK	0.86	0.37										
VOICE IN MINS	0.84											
IN CALLS OFFPEAK	0.84											
IN CALLS NONWORK	0.72											
VOICE OUT CALLS	0.47	0.85										
TOTAL OUT CALLS	0.49	0.83										
OUT CALLS OFFPEAK	0.41	0.83										
OUT CALLS PEAK	0.46	0.83										
OUT CALLS WORK	0.49	0.83										
OUT CALLS NONWORK	0.44	0.80										
VOICE OUT MINS		0.74									0.42	
OUT COMMUNITY VOICE	0.56	0.70										
OUT COMMUNITY TOTAL	0.56	0.69										
PRC OUT COMMUNITY SMS			0.92									
SMS IN CALLS			0.88									
PRC SMS OUT CALLS			0.88									
OUT COMMUNITY SMS			0.77									
PRC VOICE OUT CALLS			−0.77									
PRC IN COMMUNITY SMS			0.72									
PRC OUT COMMUNITY VOICE			−0.70									
IN COMMUNITY SMS	0.41		0.66									
SMS OUT CALLS			0.65									

Table 7.7 *(continued)*

Rotated component matrix

	Components											
	1	2	3	4	5	6	7	8	9	10	11	12
PRC IN COMMUNITY VOICE			−0.58									0.47
PRC IN CALLS OFFPEAK				0.86								
PRC IN CALLS PEAK				−0.85								
PRC IN CALLS WORK				−0.80	0.40							
PRC IN CALLS NONWORK				0.80	−0.40							
PRC OUT CALLS WORK					0.90							
PRC OUT CALLS NONWORK					−0.90							
PRC OUT CALLS OFFPEAK				0.39	−0.80							
PRC OUT CALLS PEAK				−0.38	0.79							
OUT CALLS ROAMING						0.90						
OUT MINS ROAMING						0.87						
PRC OUT CALLS ROAMING						0.86						
OUT CALLS INTERNATIONAL							0.89					
OUT MINS INTERNATIONAL							0.88					
PRC OUT CALLS INTERNATIONAL							0.86					
EVENTS CALLS								0.90				
PRC EVENTS CALLS								0.88				
EVENTS TRAFFIC								0.44				
PRC INTERNET CALLS									0.92			
INTERNET CALLS									0.90			
GPRS TRAFFIC								0.44	0.87			
PRC MMS OUT CALLS										0.92		
MMS OUT CALLS										0.91		
ACD IN											0.77	
ACD OUT											0.77	
DAYS IN	0.45	0.52										0.63
DAYS OUT												0.62

Table 7.8 Interpretation of the extracted principal components.

Component	Derived Components Label and Description
1	**Incoming voice calls** The high loadings in the first column of the rotated component matrix denote a strong positive correlation between component 1 and the original fields which measure incoming voice traffic such as the number and minutes of incoming voice calls (VOICE IN CALLS, VOICE IN MINS) and the size of the incoming voice community (IN COMMUNITY VOICE). Thus, component 1 seems to be associated with incoming voice usage Because generally voice calls constitute the majority of calls for most users and tend to dominate the total incoming usage, a set of fields associated with total incoming usage (total number of incoming calls, incoming calls by day/hour) are also loaded high on this component
2	**Outgoing voice calls** Likewise, component 2 measures the outgoing voice traffic. Therefore, users with heavy outgoing voice usage and high values in the corresponding original inputs (VOICE OUT CALLS, OUT COMMUNITY VOICE, VOICE OUT MINS, etc.) are expected to also show proportionately high values in component 2 The fact that some voice outgoing fields are also moderately loaded on component 1 implies an intercorrelation between outgoing and incoming voice calls
3	**SMS calls** Component 3 measures SMS usage, incoming as well as outgoing. The rather interesting negative correlation between component 3 and the percentage of voice out calls (PRC VOICE OUT CALLS) denotes a contrast between voice and SMS usage. Thus, users with high positive values of this component are expected to have increased SMS usage and increased SMS to voice calls ratio. This does not necessarily mean low voice traffic, but it certainly implies a relatively lower percentage of voice calls and an increased percentage of SMS calls.
4	**Day/hour of peak incoming usage** Component 4 indicates the day/hour of peak incoming usage. Users with high positive values in component 4 receive the majority of their calls in non-work days/hours. Perhaps this signifies a type of "social" use.
5	**Day/hour of peak outgoing usage** Likewise, component 5 represents the day/hour of peak outgoing usage. Users with high positive values in component 5 make most of their calls in work days/hours. Perhaps this signifies a type of "professional" use.

Table 7.8 *(continued)*

Component	Derived Components
	Label and Description
6	**Roaming usage** Component 6 measures roaming usage (making calls when abroad)
7	**International calls** Component 7 is associated with calls to fixed international networks
8	**Events and GPRS** Component 8 measures event calls and traffic. GPRS traffic is also moderately loaded high on this component
9	**Internet usage** Component 9 is associated with Internet usage and GPRS traffic
10	**MMS usage** This usage is measured by component 10
11	**ACD (Average Call Duration)** Fields denoting ACD for incoming and outgoing voice calls are related and combined to form component 11
12	**Frequency of usage** The frequency of usage, that is, the days in a month with any type of outgoing or incoming usage, is represented by component 12

The resulting 12 components retained more than 80% of the variance of the original fields. This percentage was considered more than adequate and thus the only task left before accepting the components was their interpretation. In practice only a solution consisting of meaningful components should be retained.

So what do these new composite fields represent? What business meaning do they convey? As these new fields are constructed to substitute for the original fields in the next stages of the segmentation procedure, it is necessary that they be thoroughly decoded, before being used in upcoming models.

The component interpretation phase included an examination of the rotated component matrix, a table that summarizes the correlations between the components and the original fields, that is, Table 7.7.

The "interpretation" results are summarized in Table 7.8.

The explained and labeled components and their respective scores were subsequently used as inputs to the clustering model. This brings us to the next phase of the application: the identification of useful groupings through clustering.

IDENTIFYING, INTERPRETING, AND USING SEGMENTS

The generated components represented all the usage data dimensions of interest, in a concise and comprehensible way, leaving no room for misunderstandings about their business meaning. The next step included the use of the derived component scores as inputs in a clustering model.

The clustering process involved the evaluation of different solutions obtained by trying different clustering algorithms and different model settings. All these trials resulted in overall similar, yet not identical, solutions, so it was up to the data miners and the marketers involved to select the optimal solution for deployment. The solution finally adopted was based on a TwoStep clustering model and was chosen because it seemed to best address the marketing needs of the organization.

The model automatically detected five clusters which, after extensive profiling, were assessed as meaningful and potentially useful for building the differentiated marketing strategies.

The modeling options used for the development of the adopted segmentation solution are summarized in Table 7.9.

The distribution of the derived clusters is shown in Table 7.10. These clusters were not known in advance, nor imposed by users, but were uncovered after analyzing the actual behavioral patterns recorded in the usage data.

Each revealed cluster corresponds to a distinct behavioral typology. This typology had to be understood, named, and communicated to all the people in the organization in a simple and concise way, before being used for tailored interactions and targeted marketing activities. Therefore the next phase of the project included

Table 7.9   Modeling options used to produce the segmentation solution.

Data reduction

Modeling option	*Setting*
Model	Principal components (PCA)
Rotation	Varimax
Criteria for the number of factors to extract	Eigenvalues over 1. Resulting 12 components

Clustering

Modeling option	*Setting*
Model	Two step
Input clustering fields	12 component scores derived from PCA
Number of clusters	Automatically calculated
Exclude outliers option	On
Standardize numeric fields option	On

Table 7.10 Distribution of the derived clusters.

		Percentage of MSISDNs
Clusters	Cluster 1	16.6
	Cluster 2	31.0
	Cluster 3	19.3
	Cluster 4	6.6
	Cluster 5	26.5
Total		100.0

Table 7.11 Usage KPIs by cluster.

	Clusters					
	Cluster 1	Cluster 2	Cluster 3	Cluster 4	Cluster 5	Total
VOICE OUT CALLS	32	157	35	212	235	137
VOICE IN CALLS	38	164	58	222	293	161
SMS OUT CALLS	1	21	50	15	6	19
PRC SMS OUT CALLS (%)	2	10	49	8	2	14
PRC OUT CALLS PEAK (%)	59	60	62	85	92	70
PRC OUT CALLS WORK (%)	67	70	72	88	91	77
OUT CALLS ROAMING	0.22	0.46	0.57	3.98	0.68	0.73
OUT CALLS INTERNATIONAL	0.23	0.93	0.89	2.52	0.75	0.86
MMS OUT CALLS	0.03	0.06	0.23	0.31	0.09	0.11
INTERNET CALLS	0.06	0.14	0.39	0.98	0.19	0.24
EVENTS CALLS	0.17	0.64	2.45	0.95	0.59	0.92
OUT COMMUNITY TOTAL	7	55	45	72	57	47
IN COMMUNITY TOTAL	8	52	47	56	86	53
DAYS OUT	8	25	26	27	27	23
DAYS IN	8	27	27	27	28	24
ACD OUT	1.75	1.33	0.69	2.26	0.85	1.21

the application of simple reporting techniques to profile each cluster and identify what it stands for.

This "recognize and label" procedure started with an examination of the clusters with respect to the component scores, providing a valuable first insight into their structure, before moving on to their profiling in terms of the original usage fields. The latter results are given in Table 7.11. This table summarizes the means of the majority of the original clustering fields across the derived clusters. Large deviations from the marginal mean characterize the respective cluster and denote a behavior that differentiates the cluster from the typical behavior.

Inferential statistics have also been applied to flag statistically significant differences from the overall population mean. These results (based on one-sample

Table 7.12 Cluster heat map designating statistically significant deviations from the overall (marginal) means.

	Clusters				
	Cluster 1	**Cluster 2**	**Cluster 3**	**Cluster 4**	**Cluster 5**
VOICE OUT CALLS	░		░	▓	▓
VOICE IN CALLS	░		░	▓	▓
SMS OUT CALLS	░		▓	░	
PRC SMS OUT CALLS (%)	░		▓		
PRC OUT CALLS PEAK (%)		░		▓	▓
PRC OUT CALLS WORK (%)		░			
OUT CALLS ROAMING	░			▓	
OUT CALLS INTERNATIONAL	░			▓	
MMS OUT CALLS	░		▓	░	
INTERNET CALLS	░		▓		
EVENTS CALLS	░		▓		
OUT COMMUNITY TOTAL	░			▓	
INC COMMUNITY TOTAL	░				▓
DAYS OUT	░				
DAYS IN	░				
ACD OUT	▓		░	▓	░

t-tests with a significance level of 0.01 and Bonferroni adjustment for comparisons across clusters) are illustrated in the heat map of Table 7.12. Cluster values larger than the average of the total population are represented in dark gray, and values lower than the total average in light gray.

Based on the information presented so far, the project team involved started to outline a first rough profile of each cluster:

- Cluster 1 includes users with basic usage. They are characterized by increased ACD and very low incoming and outgoing communities.
- Cluster 2 contains average voice users. They have medium SMS usage and a relatively low percentage of traffic in work days/hours.
- Cluster 3 mainly consists of SMS users who seem to have an additional inclination toward technical services like MMS, Internet, and event calls. They also present very low voice usage and ACD.

Table 7.13   Behavioral versus value-based segments.

		Clusters					
		Cluster 1	Cluster 2	Cluster 3	Cluster 4	Cluster 5	Total
Value-based segments	First class (PLATINUM Top 1%)	0	1	0	4	1	**1**
	Second class (GOLD Top 4%)	0	4	3	3	7	**4**
	Third class (SILVER High 15%)	1	19	12	15	23	**15**
	Fourth class (BRONZE Medium 40%)	10	45	50	48	43	**40**
	Fifth class (MASS Low 40%)	89	31	35	30	26	**40**
	Total (%)	100	100	100	100	100	**100**
MARPU	Mean MARPU	11	40	34	47	48	**37**
	Percentage of MSISDNs	17	31	19	7	26	**100**
	Percentage of MARPU	5	34	18	9	34	**100**

- Cluster 4 includes roamers and users with increased communication to international destinations. They are accustomed to Internet services and make most of their calls during work days/hours. They also have the largest outgoing communities.
- Cluster 5 contains heavy voice users with traffic peaks in work days/hours. Furthermore, they have the largest voice incoming communities. Perhaps this segment includes residential customers who use their phones for business purposes (self-employed professionals).

This first rough profiling of the clusters was supplemented with complementary information concerning demographic characteristics and the relationship of the derived clusters with MARPU and the value segments, already defined by the relevant segmentation.

As indicated by Table 7.13, cluster 5 includes most of the high-value users with an average MARPU of almost 48. This cluster contains about 26% of the users who account for about 34% of the total MARPU. On the other hand, cluster 1 mainly consists of mass customers who contribute a poor 5% of the total MARPU.

Table 7.14 presents the age distribution of each cluster. Cluster 1 contains the most aged part of the customer base. On the contrary, cluster 3 mainly consists of younger users. What is interesting is the relatively increased percentage of persons between 45 and 54 years old in this cluster. Perhaps this can be explained by the fact that the demographic information refers to contract holders and not users. Maybe the age data recorded upon contract registration refers to parents who

Table 7.14   The age profile of each cluster.

		Clusters					
		Cluster 1	Cluster 2	Cluster 3	Cluster 4	Cluster 5	Total
Age groups	18–24 (%)	0	1	12	1	1	**3**
	25–34 (%)	8	22	50	35	29	**28**
	35–44 (%)	13	34	17	38	33	**27**
	45–54 (%)	23	22	19	18	24	**22**
	55–64 (%)	30	14	1	5	8	**12**
	65+ (%)	26	7	1	3	5	**8**
Total (%)		100	100	100	100	100	**100**

bought the contracts for their children. Remember that the scope was to build segments based on usage and not on demographic information. Cluster assignment is based on usage patterns. Therefore, from the behavioral segmentation point of view, adults who "use" their phone in an "adolescent" way should be assigned to the relevant segment.

In order to enrich the insight into each cluster, the marketers of the organization decided to also investigate the qualitative characteristics. Therefore, they extracted a random sample from each cluster and conducted a market research survey and focus group sessions in which they examined the needs and wants and the factors that have an effect on the level of satisfaction (satisfaction drivers). The short phone interviews of the selected sample were carried out by the call center agents of the organization.

The profiling outcomes are summarized in the descriptions that follow. They present the differentiating usage characteristics, a short demographic profile, and the key market research findings on each cluster. They also present the main points of the specialized marketing approach decided for each segment. A series of summarizing charts, graphically illustrating the percentage deviations of each cluster from the overall population, are also presented as a supplement of the behavioral profile.

Segment 1: Oldies – Basic Users (Figure 7.8)

Behavioral Profile

- Lowest utilization of all services.
- Smallest outgoing and incoming communities.
- Phones used for a few days in a month.

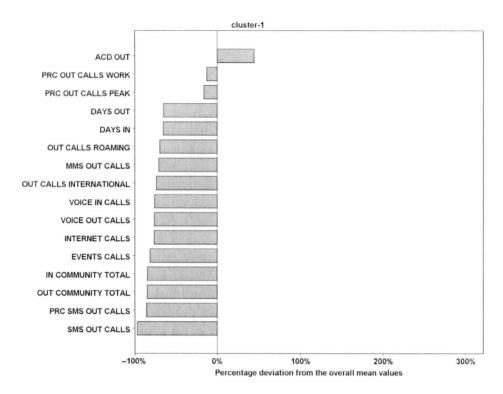

Figure 7.8 Cluster 1 profiling chart.

- When phone calls are made, however, the duration is quite long, as denoted by their ACD, which surpasses the average.

Segment Size

17%

Contribution to Total MARPU

5%

Demographic Profile

Highest ratios of elderly people, most of them retired.

Market Research Findings

- **Satisfaction drivers:**
  - Low prices.
  - Simple and easy to understand rate plans.
  - Network quality.

- **Needs/wants:**

  - Basic communication (voice) mostly with a limited number of people.
  - Easy communication.
  - Not interested in new technologies/innovative services.
  - Most calls are personal.

Marketing Approach

- **Product/services offered:** Offering of incentives to increase usage. Base rate plans with no free call time but with low monthly fees. FnF (Friends and Family) rate plans with reduced rates for selected "family" on-net numbers. Especially for elderly people, promotion of simple and easy-to-use handset devices
- **Communication channel:** Direct voice calls, direct mail.
- **Brand image communicated:** An established, trustworthy, and reliable operator.

Segment 2: Adults – Classic Social Users (Figure 7.9)

Behavioral Profile

- Average usage.
- Mostly voice usage.
- Medium SMS usage.
- Low usage of technical services (MMS, events, Internet).
- Average communities and days with usage.
- Relatively increased usage during no work days/hours.

Segment Size

31%

Contribution to Total MARPU

34%

Demographic Profile

Typical age profile, mainly consisting of married users.

Market Research Findings

- **Satisfaction drivers:**

  - Competitive pricing.

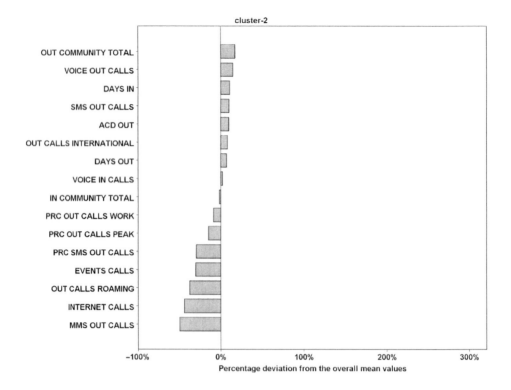

Figure 7.9 Cluster 2 profiling chart.

- Optimum rate plan.
- Reliable services.
- **Needs/wants:**

  - Reliable communication, mostly with friends and family.
  - Seek value for money.
  - Most calls are personal.

Marketing Approach

- **Product/services offered:** Offering incentives to promote use of non-voice services. Optimized rate plans, tailored to their usage, providing free voice minutes or reduced rates for voice calls.
- **Communication channel:** Direct voice calls.
- **Brand image communicated:** An operator that offers reliable services at competitive prices.

Segment 3: Young – SMS Users (Figure 7.10)

Behavioral Profile

- Heaviest SMS usage, almost three times above the average.
- Low voice usage.
- Mostly use handsets for sending SMS messages. By far the highest percentage of SMS calls and the lowest percentage of voice calls.
- Voice usage is predominantly incoming (perhaps as a result of receiving calls from worried parents!).
- Voice calls are brief, as denoted by their ACD, which is the lowest among all segments.
- More than comfortable with the new technologies, especially with technical services such as MMS and events.
- Internet usage is the second highest.

Segment Size

19%

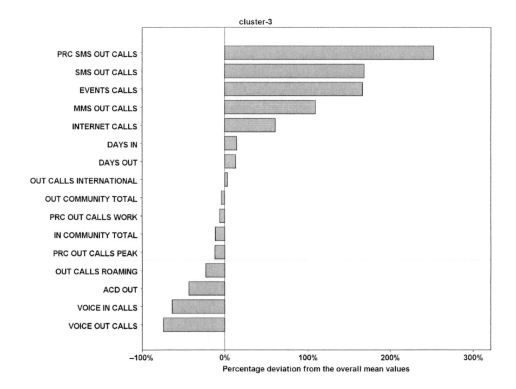

Figure 7.10   Cluster 3 profiling chart.

Contribution to Total MARPU

18%

Demographic Profile

Young persons, most of them under 35. Predominantly students, young employees, and single people.

Market Research Findings

- **Satisfaction drivers:**

  - Offering a wide range of innovative services.
  - Competitive pricing, especially for SMS.
  - "Cool" brand image.

- **Needs/wants:**

  - Interested in new technologies/innovative services.
  - Use of phone not only for communication but also for fun.
  - Identify with the phone, so it must be "cool."
  - Seek and compare competitors' offers.
  - Most calls are with friends and family.

Marketing Approach

- **Product/services offered:** Offering of "Bring a friend" incentives. Promotion of rate plans and bundles with SMS and MMS discounts. Promotion of cell phones with advanced technical capabilities. Promotion of operator's web site and e-billing services. This is the ideal target group for promoting all high-tech services: event downloads (ringtones, games, MP3s), Internet, mobile TV, video calls, and so on.
- **Communication channel:** SMS, MMS, e-mails.
- **Advertising channels:** Internet banners, TV channels, radio stations, and magazines influential to young people.
- **Brand image communicated:** Cool and stylish brand providing fun/ entertainment and innovative services.

Segment 4: International Users (Figure 7.11)

Behavioral Profile

- Citizens of the world! Seem to travel a lot so show the highest roaming usage.
- Highest number of roaming calls and roaming traffic minutes.

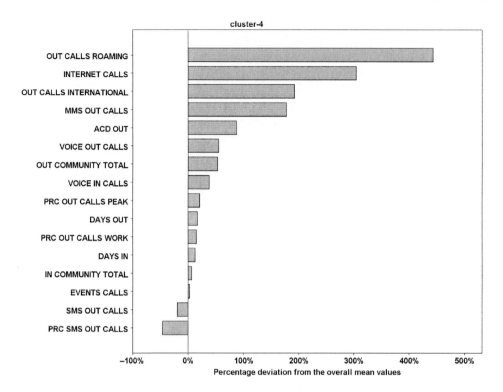

Figure 7.11 Cluster 4 profiling chart.

- Frequent calling of international numbers. Have the highest number of international calls and minutes.
- Most calls made during work days/hours.
- Accustomed to the new types of communication, with high Internet, events, and MMS usage.
- Top users, particularly in terms of the Internet. Probably use it a lot when traveling abroad.
- Largest outgoing community and with the highest ACD.
- Voice usage is also high.

Segment Size

7%

Contribution to Total MARPU

9%

Demographic Profile

Young and middle-aged persons, most of them under 45. This cluster mainly consists of male employees and self-employed people with a college degree.

Market Research Findings

- **Satisfaction drivers:**

  - Network coverage.
  - Efficiency of customer care.
  - Roaming quality.
- **Needs/wants:**

  - The phone is a business tool that should provide support around the world.
  - Interested in new technologies/innovative services.
  - "Always connected."

Marketing Approach

- **Product/services offered:** Offering of rate plans and bundles with roaming and international call benefits. Teaming up with airlines and travel agencies, in order to offer "air miles and more" reward schemes with point offerings according to phone usage. Promotion of company's web site and Internet usage.
- **Communication channel:** Direct voice calls, e-mails, MMS.
- **Advertising channels:** Internet banners, news channels and stations, newspapers, travel magazines, airport advertisements.
- **Brand image communicated:** A modern operator, with a presence and reliability around the world, which offers a wide range of innovative services.

Segment 5: Business Users (Figure 7.12)

Behavioral Profile

- Heavy voice usage, incoming as well as outgoing.
- Majority of calls made during work days and peak hours.
- By far the largest incoming community.
- Highest number of voice calls and minutes. This segment includes many residential customers using their phone as a business tool as well.
- Although the phone is used on a daily basis, voice calls are quite short, as denoted by the relatively low ACD.
- Low SMS usage.

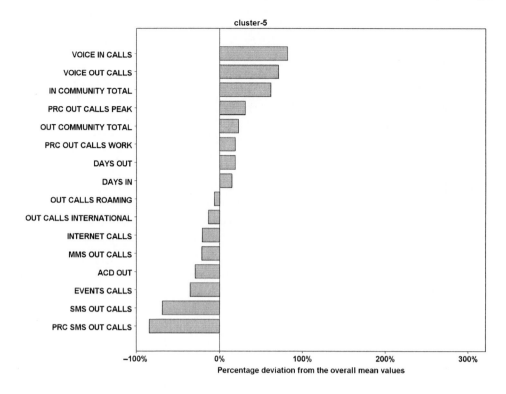

Figure 7.12   Cluster 5 profiling chart.

Segment Size

26%

Contribution to Total MARPU

34%

Demographic Profile

Middle-aged persons. Mainly includes self-employed professionals and employees in high-level positions.

Market Research Findings

- **Satisfaction drivers:**

    - Network coverage.
    - Reliable services.
    - Efficiency of customer care.

  - Competitive pricing.
  - Optimum rate plan.
- **Needs/wants:**

  - Phone used as a business tool.
  - High-quality services and guaranteed availability at competitive prices.
  - "Always connected."

Marketing Approach

- **Product/services offered:** Offering of business rate plans with free minutes and discounts on voice usage. Offering of incentives for handset replacement.
- **Communication channel:** Direct voice calls, direct mail.
- **Advertising channels:** News channels and stations, newspapers, business magazines.
- **Brand image communicated:** Trustworthy, high-status company, the choice of successful people.

SEGMENTATION DEPLOYMENT

The final stage of the segmentation process involved the integration of the derived scheme into the company's daily business procedures. From that moment all users were characterized by the segment to which they were assigned. The deployment procedure involved a regular (on a monthly basis) segment update for the needs of which a specialized classification model was developed. More specifically, a decision tree model (C5.0) was trained, using all the initial clustering fields as inputs and the segment membership field as the target. The generated model was used not only to gain insight into the segments, but also as a scoring engine for allocating new records to the established clusters. Although the generated cluster model has been saved and could also be used for scoring, the decision tree solution was preferred due to its transparency. Decision trees translate the distinct profiles of the segments into a set of intuitive rules, similar to common business rules. Therefore they facilitate a clear understanding of the defining characteristics of each cluster and enable simpler communication within the organization. Moreover, they are easier to handle, allowing possible interventions in and modifications to the assignment procedure, if considered necessary.

A part of the resulting decision tree is shown in Figure 7.13. Although only the top two levels of the tree are presented here, the beginning of segment separation is evident. Additionally, a subset of the tree's rules is shown in Figure 7.14.

Classification rules were identified for all segments. All records passing through the generated model are allocated by these rules to the segment that best

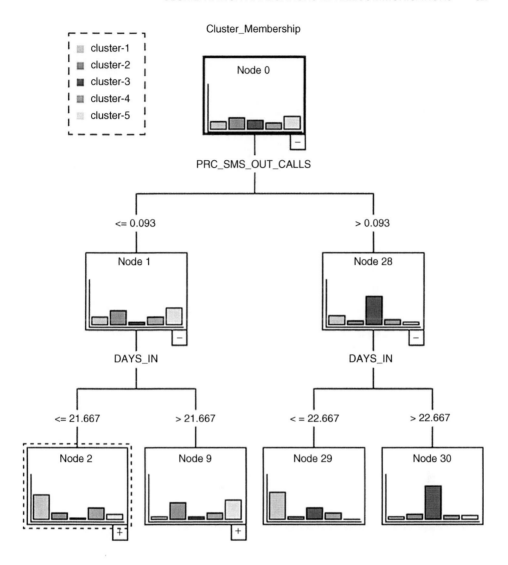

Figure 7.13   Use of a decision tree for profiling and scoring segments.

matches their behavior. For example, we notice that users with a relatively high percentage of SMS usage (above 10%) and an increased frequency of inbound usage are assigned to segment 3 as "SMSers." Similarly, records with a high ratio of outgoing voice calls (above 90%), high frequency of both incoming and outgoing usage (higher than 22 and 29 days respectively), and high number of outgoing calls in peak hours (59 and above) land in segment 5 as "Business users."

Subsequent steps in the deployment included the loading of the derived segment information into the operational and front-line systems. Segmentation

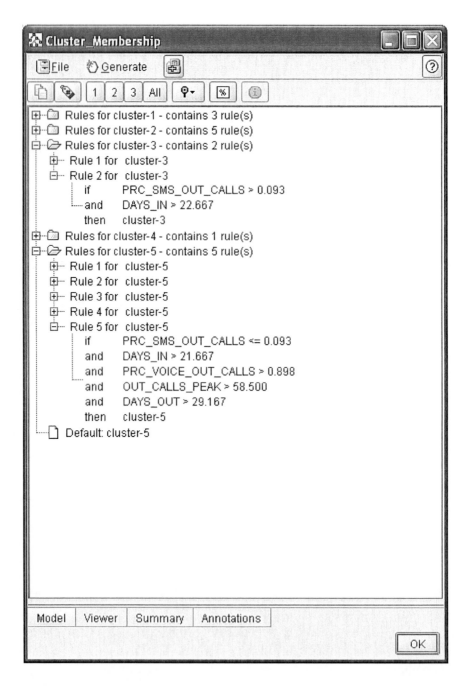

Figure 7.14 A subset of the rule set developed for profiling and scoring segments.

information and guidelines for the specialized handling of each customer according to the profile of its segment were made available to all points of service.

THE FIXED TELEPHONY CASE

Nowadays, fixed phone operators are offering much more than plain voice services. They provide a full range of telecommunication products, from Internet access and broadband services to Internet Protocol TV (IPTV) services. The diverse wants and needs of the customers are also reflected in different usage behaviors. Fortunately these differentiations are recorded in traffic data which can be mined to reveal the distinct customer typologies. In times when the market changes dramatically, offering new challenges, new possibilities, and new fields for development, a thorough understanding of an evolving customer base is necessary and the role of segmentation is vital.

The behavioral segmentation procedure in fixed telephony is in many ways similar to the mobile telephony case which was described in detail in the previous sections of this chapter. It aims at identifying natural groupings of customers according to their usage patterns and utilization of the services offered. The main data dimensions which should be covered by such an analysis are relevant to the ones presented for mobile telephony and include:

- Product/service utilization and usage traffic:

  - Voice usage
  - Internet usage
  - Broadband usage
  - Digital TV usage
  - Voice mail usage, and so on.

- Community size:

  - Outgoing community
  - Incoming community.

- Usage by origination/destination network:

  - Long-distance calls
  - International calls
  - On-net calls
  - Off-net calls, and so on.

- Usage by peak/off-peak days and hours:

  - Peak/off-peak hours calls
  - Work/non-work days calls.

- Tenure.

  Customer typologies expected to be found include:

- Young families:

  - New connections
  - Ages between 27 and 35
  - Relatively high broadband penetration.

- Basic users:

  - Low usage
  - Price sensitive, low ACD
  - Ages between 35 and 55
  - Classic voice usage only.

- Classic users:

  - Normal/average usage
  - Ages between 35 and 55.

- Professionals:

  - High usage at peak hours
  - Large incoming community
  - Use of voice mail and broadband
  - International calls.

- Superstars:

  - Very high levels of usage at off-peak hours
  - Highest broadband penetration
  - High ACD
  - International calls
  - Digital TV and video-on-demand packs.

- Seniors:

  - Basic voice usage
  - Low incoming community
  - Ages over 60.

SUMMARY

In this chapter we have followed the efforts of a mobile telephony network operator to segment its customers according to their usage patterns and their value. The business objective was to group customers in terms of their profit and their behavioral characteristics and to use this insight to deliver personalized and value-driven customer handling. The organization initially focused on residential postpaid customers, who were further segregated into five behavioral and five value-based segments, as summarized in Table 7.15.

Table 7.15   The mobile telephony segments.

Residential customers	Postpaid – contractual	
	Value-based segments	*Behavioral segments*
	Platinum	Oldies – basic users
	Gold	Adults – classic social users
	Silver	Young – SMS users
	Bronze	International users
	Mass	Business users
	Prepaid	
Business customers	**Large business – corporate**	
	Small and medium enterprise (SME)	
	Small office, home office (SOHO)	

The procedure followed for the behavioral segmentation included the application of a PCA model for data reduction and a cluster model for revealing distinct user groups. Furthermore, a decision tree model was also applied to profile the segments and compile a set of understandable scoring rules for updating the segments. The whole process is depicted in Figure 7.15.

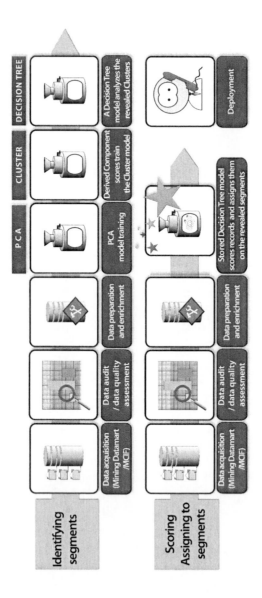

Figure 7.15 The segmentation building and scoring procedure.

CHAPTER EIGHT

Segmentation for Retailers

SEGMENTATION IN THE RETAIL INDUSTRY

Retail enterprises offer a large variety of products through different channels to customers with diverse needs. The lack of a formal commitment and the ease with which shoppers can prefer competitors make the process of building a loyal customer base tougher for retailers. Quality of commodities and competitive pricing are often not enough to stand out from the competition.

In such a dynamic environment a competitive enterprise should try to understand its customers by gaining insight into their needs, attitudes, and behaviors. Ideally each customer should be treated as an individual and the enterprise should operate on a one-to-one basis. Since this approach is obviously not possible, an efficient alternative is to segment the customers into groups with different characteristics and develop differentiated strategies that best address their specific features.

As mentioned in a previous chapter, different segmentation schemes can be developed according to the specific business objectives of the organization.

Needs/attitudinal segmentation is commonly employed through market research data in the retail industry to gain insight into the customer attitudes, wants, views, preferences, and opinions about the enterprise and the competition. In addition to external/market research data, transactional data can also be used for the development of effective segmentation solutions. A value-based segmentation scheme allocates customers to groups according to their spending amount. It can be used to identify high-value customers and to prioritize their handling according to their measured importance.

Data Mining Techniques in CRM: Inside Customer Segmentation   K. Tsiptsis and A. Chorianopoulos
© 2009 John Wiley & Sons, Ltd

Behavioral segmentation is also based on transactional data and separates customers according to attributes that summarize their shopping habits, such as:

- Frequency and recency of purchases
- Total spending amount
- Relative spending amount per product group/subgroup
- Size of basket (spending amount and number of items per visit or transaction)
- Preferred payment method
- Preferred period/day/time of purchases
- Preferred store and channel, and so on.

The derived segments can be used for the "personalized" handling of segmented customers through the development of differentiated sales and marketing strategies, tailored to their recognized consuming habits.

Transactional data are logged at the point of sale and typically record the detailed information of every transaction, including the universal product code (UPC) of each purchased item, which allows detailed monitoring of the groups and subgroups of products that each customer tends to buy. A prerequisite of behavioral segmentation is that every transaction is identified with a customer. This issue is usually tackled by introducing a loyalty program which assigns an identification field (card ID) to each transaction and permits the tracking of the purchase history of each customer and aggregation of the transactional information at a customer level.

In this chapter we focus on the efforts of a retailer to segment its customers according to their consuming habits and more specifically according to the product mix they buy. A high-level grouping of products was selected for this first segmentation attempt. The relevant data were readily available within the organization's mining data mart and MCIF which stored all the processed transactional information. In addition, the marketers of the organization also decided to employ a recency, frequency, monetary (RFM) analysis to examine and group their customers according to their purchase frequency, recency, and value. These applications are described in the following sections.

THE RFM ANALYSIS

RFM analysis is a common approach for understanding customer purchase behavior. It is quite popular, especially in the retail industry. As its name implies, it involves the calculation and the examination of three KPIs – recency, frequency, and monetary – that summarize the corresponding dimensions of the customer relationship with the organization. The recency measurement indicates the time since the last purchase transaction of the customer. Frequency denotes the number and rate of purchase transactions. Monetary indicates the value of the purchases. These indicators are typically calculated at a customer (card ID) level through simple data processing of the available transactional data.

RFM analysis can be used to identify good customers with the best scores in the relevant KPIs, who generally tend to be good prospects for additional purchases. It can also identify other purchasing patterns and respective customer types of interest, such as infrequent big-spenders or customers with small but frequent purchases who might also have sales perspectives, depending on the market and the specific product promoted.

In the retail industry, the RFM dimensions are usually defined as follows:

- **Recency:** The time (in units such as days/months/years) since the most recent purchase transaction or shopping visit.
- **Frequency:** The total number of purchase transactions or shopping visits in the period examined. An alternative, and probably a better defined, approach that also takes into account the tenure of the customer calculates frequency as the average number of transactions per unit of time, for instance the monthly average number of transactions.
- **Monetary:** The total value of the purchases within the period examined or the average value (e.g., monthly average value) per time unit. According to an alternative, but not so popular, definition, the monetary indicator is defined as the average transaction value (average value per purchase transaction). Since the total value tends to be correlated with the frequency of the transactions, the reasoning behind this alternative definition is to capture a different and supplementary aspect of purchase behavior.

The construction of the RFM indicators is a simple data management task which does not involve any data mining modeling. It does, however, involve a series of aggregations and simple computations that transform the raw purchase records into meaningful scores. In order to perform RFM analysis each transaction should be linked with a specific customer (card ID) so that the customer's purchase history can be tracked and investigated over time. Fortunately, in most situations, the use of a loyalty program makes the collection of "personalized" transactional data possible.

RFM components should be calculated on a regular basis and stored along with the other behavioral indicators in the organization's mining data mart and MCIF table. They can be used as individual fields in subsequent tasks, for instance as inputs, along with other predictors, in upcoming supervised cross-selling models. They can also be included as clustering fields for the development of a multi-attribute behavioral segmentation scheme. Usually they are simply combined to form a single RFM measure and a respective cell-based segmentation scheme.

Typically, RFM analysis involves the grouping (binning) of customers into chunks of equal size, or quantiles, in a way similar to the one presented for value-based segmentation. This binning procedure is applied independently to the three RFM component measures. Customers are sorted according to the respective measure and are grouped in classes of equal size. For instance, the breakdown into four groups results in quartiles of 25%, and the breakdown into five groups, in quintiles of 20%. As a consequence the RFM measures are transformed into ordinal scores. In the case of binning into quintiles, for example, the RFM measures are converted into rank scores ranging from 1 to 5. Group 1 includes the 20% of customers with the lowest values and group 5 the 20% of customers with the top values in the corresponding measure. Especially for the recency measure, the scale of the derived ordinal score should be reversed so that larger scores represent the most recent buyers.

The derived R, F, and M bins become the components for the RFM cell assignment. These bins are combined with a simple concatenation to provide the cell assignment. Customers with the top RFM values and quintile values of 5 are assigned to cell 555. Similarly, customers with the average recency (quintile 3), top frequency (quintile 5), and lowest monetary values (quintile 1) form cell 351, and so on.

This procedure for constructing the RFM cells is illustrated in Figure 8.1.

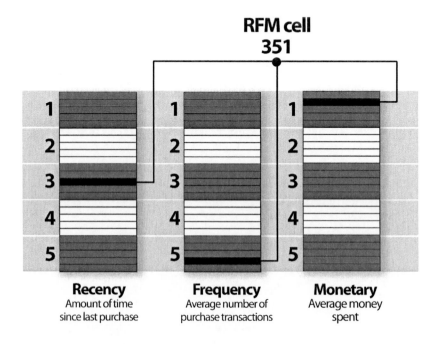

Figure 8.1   Assignment to the RFM cells.

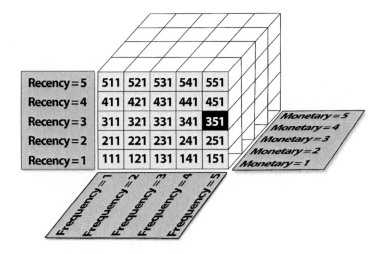

Figure 8.2   The total RFM cells in the case of binning into quintiles (groups of 20%).

When grouping customers in quintiles (groups of 20%), the procedure results in a total of $5 \times 5 \times 5 = 125$ RFM cells as displayed in Figure 8.2.

This combination of the R, F, and M components into cells is widely used, though it does have a certain disadvantage. The large number of derived cells makes the procedure quite cumbersome and hard to manage. An alternative method for segmenting customers according to their RFM patterns is to use the respective components as inputs in a clustering model and let the algorithm reveal the underlying natural groupings of customers.

The marketers of the retail enterprise decided to perform RFM analysis, before proceeding to the development of a more general multi-attribute segmentation scheme. The procedure followed is presented in "The RFM Segmentation Procedure".

Combining R, F, and M Components to Derive a Continuous RFM Score

An alternative approach treats the binned R, F, and M components as continuous measures. According to this approach, the R, F, and M bins are summed, with appropriate user-defined weights, in order to provide a continuous RFM score. The RFM score is the weighted average of its individual components and is calculated as follows:

RFM score = (recency score × recency weight)
            + (frequency score × frequency weight)
            + (monetary score × monetary weight).

The weights assigned to each RFM component designate its significance and can be specified by the analysts according to prior knowledge of the particular industry and enterprise.

As an example let us consider the case of the customer previously assigned to RFM cell 351. With equal weights of 10.0 in all the RFM individual components, this customer would receive a score of 90, in a scale ranging from 30 to 150. These scores can be rescaled to the 0–1 range according to the following formula:

$$\text{Rescaled RFM score} = \frac{\text{RFM score–minimum RFM score}}{\text{maximum RFM score–minimum RFM score}}.$$

Unlike the RFM cells, the continuous form of the RFM unifies the information in the relevant components without preserving their distinct information. Nevertheless, this can prove to be simpler and more convenient since marketers will have to monitor just a single scale measurement instead of deciphering a rather complex combination of numbers.

THE RFM SEGMENTATION PROCEDURE

Since the goal was to investigate current and not past behaviors, the RFM analysis was based on purchase records of the last six months. The procedure involved the calculation of the RFM components and the assignment of customers to RFM segments. The implementation steps for the calculation of the RFM components and cells are briefly as follows:

1. **Data acquisition:** Transactional data of the last six months were retrieved, audited, cleaned, and prepared for subsequent operations. The organization's loyalty program made it possible to link customers to specific transactions and to track each customer's purchase trail.
2. **Selection of the population to be segmented:** Only active customers were included in the RFM segmentation. Inactive customers with no purchases during the last six months were identified and a relevant list was constructed and sent to the marketing department for inclusion in upcoming retention and reactivation campaigns.

   Furthermore, new customers with a relationship with the retail enterprise of less than three months were also excluded from the RFM analysis since it was considered that their relationship with the enterprise was relatively short and the available data were not sufficient to outline a reliable purchase profile.

   Finally, after preliminary data analysis and discussions with the marketers of the organization, customers with total purchases lower than a specific threshold value were considered as dormant and were also excluded from the subsequent segmentation.

3. **Data preparation and computation of the R, F, and M measurements:** Transactional data were aggregated (grouped by) at a customer (card ID) level.

A two-fold aggregation procedure was followed. Initially, records were grouped by card ID and transaction ID. Then they were further grouped by card ID as shown in the IBM SPSS Modeler Aggregate screenshot in Figure 8.3. The information summarized for each customer included:

(a) Date of the latest (maximum in the case of a date or timestamp field) purchase transaction. This information was then used to derive recency as the number of days since the most recent purchase transaction. The IBM SPSS Modeler Derive node and a date function were used to return the number of days from the last transaction to the current date (represented

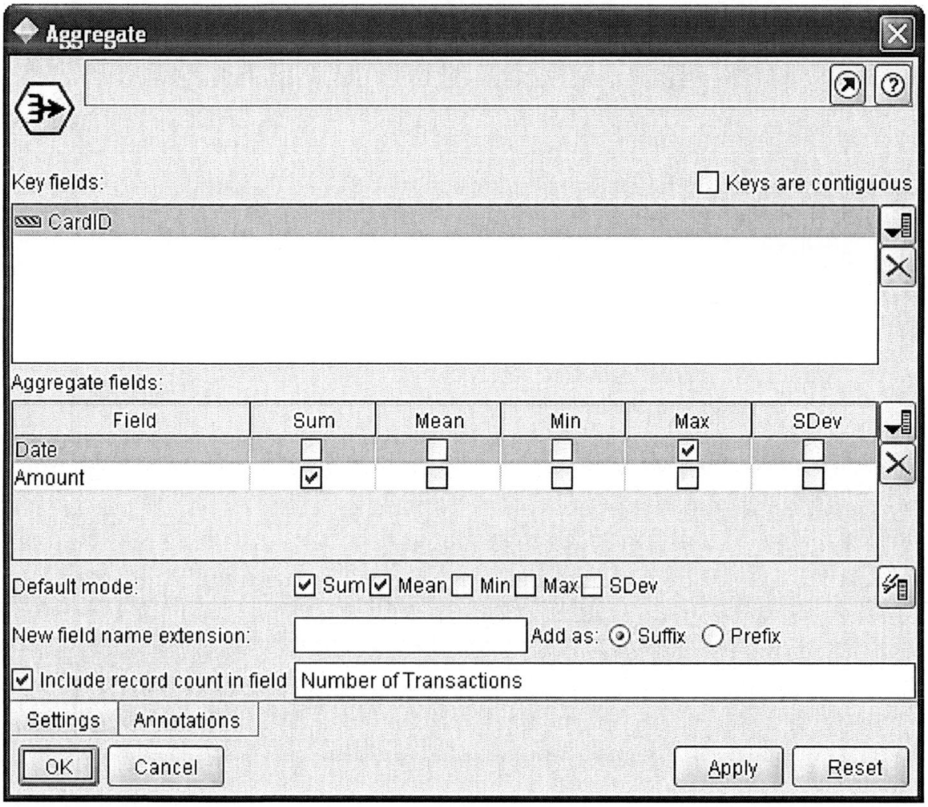

Figure 8.3 The IBM SPSS Modeler Aggregate node for summarizing purchase data at a customer level.

by IBM SPSS Modeler's "@TODAY" function), as displayed in the screen-shot in Figure 8.4.

(b) Monthly average number of distinct purchase transactions. This information defined the frequency component of the RFM analysis. As shown in Figure 8.5, a conditional derive node in IBM SPSS Modeler was used to divide the total number of transactions by the appropriate number

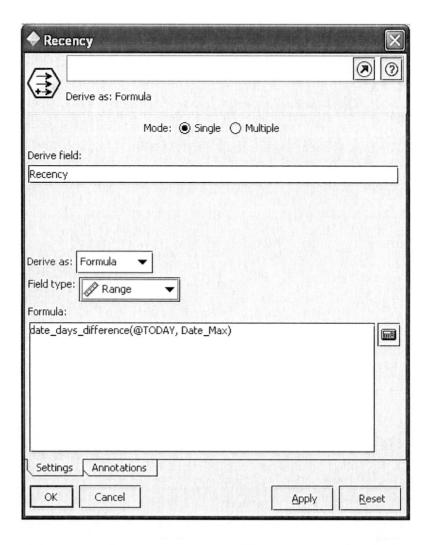

Figure 8.4 Deriving the recency component of the RFM score with a date function in IBM SPSS Modeler.

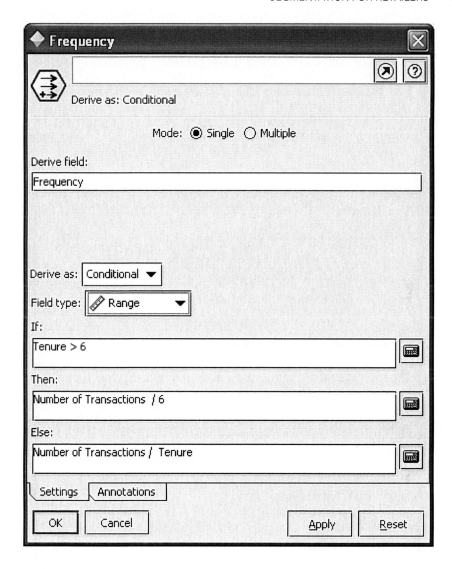

Figure 8.5 Deriving the frequency component with a conditional derive node in IBM SPSS Modeler.

of months: six months for old customers and the number of months as registered customers (tenure) for new customers.

(c) Monthly average amount spent defined the monetary component. This component was calculated with a formula similar to the one used for the frequency measure.

The IBM SPSS Modeler (Formerly Clementine) RFM Aggregate Node

IBM SPSS Modeler also includes a data preparation tool, named the "RFM Aggregate" node (Figure 8.6), that can simplify the computation of the individual RFM components. Users only have to designate the required source fields, specifically the customer ID (card ID) field for the aggregation and the fields indicating each transaction's value and date.

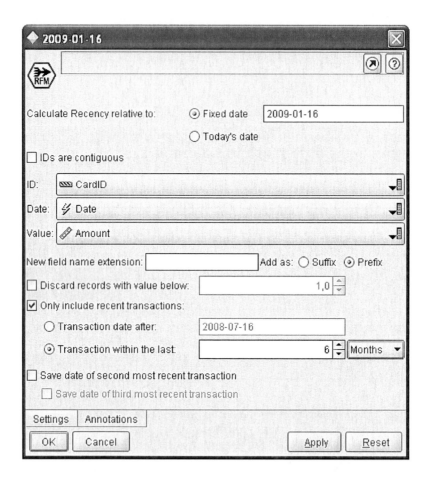

Figure 8.6   The IBM SPSS Modeler RFM Aggregate node.

4. **Development of the RFM cells through binning:** Customers were sorted independently according to each of the individual RFM components and then

binned into five groups of 20%. The resulting bins (quintiles) were then combined (concatenated) to form the RFM cell assignment.

The IBM SPSS Modeler (Formerly Clementine) RFM Analysis node

IBM SPSS Modeler also offers a tool, named the "RFM Analysis" node, that can directly group the R, F, and M measures into the selected number of quantiles. Users can then collate the derived ordinal scores to form the corresponding cell-based segmentation. Additionally, this node also sums the individual components by using user-defined weights to produce a continuous RFM score. The IBM SPSS Modeler "RFM Analysis" node is illustrated in Figure 8.7.

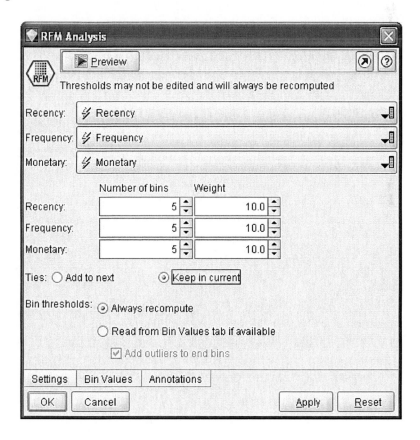

Figure 8.7 The IBM SPSS Modeler RFM Analysis node.

5. **Development of the RFM segments through clustering:** The marketers of the enterprise decided to also apply a clustering model to analyze the RFM components. They also decided to enrich the segmentation criteria with information concerning the purchases of private label products and the average basket value. This value denotes the average amount spent on each transaction. Private label products are goods which are only sold in the retailer's stores under the retailer's brand name and are often positioned as low-cost alternatives to "named" brands. The retailer was interested in investigating the favoring of these special products. Therefore the percentage (%) of the total purchase amount accounted for by private label products was included in the clustering fields along with the RFM individual components and the basket value. The clustering model identified six RFM segments, which are listed in Table 8.1 along with their identified characteristics.

The majority of customers with average RFM patterns were assigned to the "Typical" segment. The "Dormant" segment contained customers on the verge of inactivity with very low purchase rates and the worst RFM profile. At the other end stood the "Superstars" and the "Golden customers". These were high-value customers with increased frequency of transactions. "Superstars" in particular seemed to be the most loyal ones, with an increased number of visits/transactions and a high preference for private label products. "Everyday

Table 8.1   The RFM segments revealed through clustering.

Superstars	• The most loyal customers • Highest value • Highest frequency • High spending on private labels
Golden customers	• Second highest value • High frequency • Average spending on private labels
Typical customers	• Average value and frequency • Average spending on private labels
"Exceptional occasions" customers	• The second lowest frequency after "Dormant customers" • Large basket • Low recency values (long time since their last visit)
"Everyday" shoppers	• Increased frequency of transactions • Small basket • Private labels • Medium to low value
Dormant customers	• Lowest frequency and value • Long time since their last visit (lowest recency values)

shoppers" made frequent but low-value transactions, probably to cover their daily needs. They also showed increased preference for private label brands. Occasional customers on the other hand made infrequent visits to the store branches but of high average value.

A deployment procedure was also developed to support future updating of the segments. The clustering model was supplemented with a classification model, a decision tree in particular, which identified the input patterns associated with each revealed RFM segment. The tree rules were saved for the segment assignment of new records. The deployment plan also included a periodic cohort analysis, a type of "before–after" examination of the customer base, with simple reports that could identify the migrations of customers across segments over time.

RFM: BENEFITS, USAGE, AND LIMITATIONS

The individual RFM components and the derived segments convey useful information with respect to the purchasing habits of consumers. Undoubtedly, any retail enterprise should monitor the purchase frequency, intensity, and recency as they represent significant dimensions of the customer's relationship with the enterprise.

Moreover, by following RFM transitions over time an organization can keep track of changes in the purchasing habits of each customer and use this information to proactively trigger appropriate marketing actions. For instance, specific events, like the decline in the total value of purchases, a sudden drop in the frequency of visits, or no-shows for an unusually long period of time, may indicate the beginning of the end of the relationship with the organization. These signals, if recognized in time, should initiate event-triggered retention and reactivation campaigns.

RFM analysis was originally developed for retailers, but with proper modifications it can also be applied in other industries. It originated from the catalogue industry in the 1980s and proved quite useful in targeting the right customers in direct marketing campaigns. The response rates of the RFM cells in past campaigns were recorded and the best-performing cells were targeted in the next campaigns. An obvious drawback of this approach is that it usually ends up with almost the same target list of good customers, who could become annoyed with repeated contacts. Although useful, the RFM approach, when not combined with other important customer attributes such as product preferences, fails to provide a complete understanding of customer behavior. An enterprise should have a complete view of the customer and use all the available information to guide its business decisions.

These limitations led the marketers of the organization to develop an additional segmentation scheme that would separate customers with respect to the products that they tend to buy. Their purchases by product category were analyzed and the customers were grouped accordingly. This procedure is outlined in the next section. The revealed segments also reflected the lifecycle stage of the customers and provided valuable information for the development of tailored marketing activities.

GROUPING CUSTOMERS ACCORDING TO THE PRODUCTS THEY BUY

The next step was to reveal the customer types with respect to the mix of products that they tend to buy. The objective was to use the results to optimize and adapt the offers, rewards, and incentives received by customers according to their identified needs and preferences. The segmentation process involved the application of a clustering model to the purchase records. Relevant data from the last six months were aggregated at a customer (card ID) level. A high hierarchy level in the existing product taxonomy was chosen for grouping the product codes, as follows:

- Apparel/shoes/jewelry
- Baby
- Electronics
- Computers
- Food and wine
- Health and beauty
- Pharmacy
- Sports and outdoors
- Books/press
- Music/movies/videogames
- Toys
- Home.

The segmentation fields summarized the relative spending (percentage of total spending) of each active customer in the above product categories. Demographic data, including age, gender, and marital status of the customers, were not included in the model training; however, they contributed to the profiling of the clusters generated. The revealed segments are presented in Table 8.2 along with a brief behavioral and demographic profile.

Customers were assigned to six groups. Although the segmentation criteria only involved purchasing preferences, the demographic profile of the clusters also

Table 8.2 Customer segments with respect to the product mix.

Segment	Purchase patterns	Demographics
1. Hobby shoppers – single men	Electronics, computers, movies/music/games, sports and outdoors	Young men, single
2. Fashion shoppers – single women	Apparel/shoes/jewelry, beauty, books/press, movies/music/games	Young women, single
3. Average shoppers	Home, food and wine, apparel/shoes/jewelry, electronics, movies/music/games, health and beauty	Young, married, no children
4. Family shoppers – full nest I	Food and wine, home, baby, toys	Young, married, children under 5
5. Family shoppers with children – full nest II	Food and wine, home, toys, books/press, movies/music/games	Middle aged, married with children aged 5 or over
6. Older families/retired	Home, food and wine, pharmacy, books/press	Older, no dependents

revealed a clear separation between age and marital status. This finding confirmed the initial belief that the product mix would be strongly associated with the family lifecycle stage of the customers.

The first cluster was dominated by single young men. "Hobby shoppers" were characterized by increased relative spending on recreation products such as electronics, computers, movies/music/games, and sports and outdoors products, revealing a behavior typical of lively young men with free time and many leisure activities. "Fashion shoppers," the female counterpart of the first cluster, showed a similar purchasing profile, with an increased preference for clothes, accessories, and beauty products.

The segment of "Average shoppers" has diverse preferences and spending on many product categories, including grocery, home, and leisure products. Young, married customers formed the majority in this cluster. It seems that customers at this life-stage start to spend on their home and family; however, they still spend money on themselves and their hobbies.

Different needs and new priorities arise with the birth of a baby. Customers with high spending on baby products formed a group of their own, the segment of "Family shoppers," mainly consisting of young married people with a baby. As babies grow into children, the purchase baskets change once again. Grocery and home products can still be found in the family's basket but the baby products are replaced by toys, books, movies/music/games; that is, leisure time products, but for the children this time, not for the parents. This purchasing profile characterized the segment of "Family shoppers with children," a group mainly composed of middle-aged married customers. The last cluster included customers with low total purchases, mainly for home, food, and pharmacy products. It was labeled as "Older families/retired" since it turned out to mostly contain older people with no dependents.

Once the product-based segments had been revealed and their differentiating characteristics recognized, it was time for the marketers at the retailer to act on them and use them to customize the loyalty program's offerings. From that point on, customers were presented with rewards, offers, and incentives that matched their specific profile. The reward type was determined by the identified customer profile, but its value depended on the RFM segment of the customer.

Finally, the enterprise also decided to differentiate its communication strategy, according to the identified customer typologies. An initiative in this direction included the replacement of general interest brochures and newsletters with specialized but more detailed ones which better addressed the specific requirements of each revealed segment.

SUMMARY

In this chapter we presented a segmentation example from the retail industry. The retailer's loyalty program made possible the tracking of each customer's purchases over time. Purchase data of the last six months were mined, revealing the purchasing preferences of the customers and enabling their product-based segmentation. A clustering model revealed six segments, clearly differentiated in terms of their consuming preferences. The purchasing patterns identified also reflected the different family lifecycle stages of the customers, their differentiated needs, wants, and priorities. Moreover, this allowed the marketers of the organization to customize their loyalty program's offers and rewards and the communication strategies of the enterprise, according to the distinct profile of each segment.

Customers were also segmented according to their purchase recency, frequency, and money. This RFM segmentation was also taken into account in prioritizing the customer handling according to the importance of each customer.

FURTHER READING

Anderson, Kristin. *Customer Relationship Management*. New York: McGraw-Hill, 2002.

Berry, Michael J. A., and Gordon S. Linoff. *Data Mining Techniques For Marketing, Sales, and Customer Relationship Management*. Wiley Computer, 2004.

Berry, Michael J. A., and Gordon S. Linoff. *Mastering Data Mining: The Art and Science of Customer Relationship Management*. New York: Wiley, 1999.

Fernandez, George. *Data Mining Using SAS Applications*. Boca Raton: Chapman & Hall/CRC, 2003.

IBM Redbooks. *Mining Your Own Business in Banking Using DB2 Intelligent Miner for Data* (IBM Redbooks). IBM, 2001. http://www.redbooks.ibm.com/.

IBM Redbooks. *Mining Your Own Business in Retail Using DB2 Intelligent Miner for Data*. IBM, 2001. http://www.redbooks.ibm.com/.

IBM Redbooks. *Mining Your Own Business in Telecoms Using DB2 Intelligent Miner for Data (IBM Redbooks)*. IBM, 2001. http://www.redbooks.ibm.com/.

Kimball, Ralph, and Margy Ross. *The Data Warehouse Toolkit: The Complete Guide to Dimensional Modeling (Second Edition)*. New York: Wiley, 2002.

Kotler, Philip, and Kevin Lane Keller. *Marketing Management (12th Edition) (Marketing Management)*. Upper Saddle River: Prentice Hall, 2005.

Larose, Daniel T. *Discovering Knowledge in Data: An Introduction to Data Mining*. Hoboken, N.J: Wiley-Interscience, 2005.

Linoff, Gordon S. *Data Analysis Using SQL and Excel*. New York: Wiley, 2007.

Matignon, Randall. *Data Mining Using SAS Enterprise Miner*. Hoboken, N.J: John Wiley, 2007.

Olson, David, and Dursun Delen. *Advanced Data Mining Techniques*. Berlin Heidelberg: Springer-Verlag, 2008.

Peelen, Ed. *Customer Relationship Management*. Upper Saddle River: Financial Times/Prentice Hall, 2005.

Peter, Chapman, Julian Clinton, Randy Kerber, Thomas Khabaza, Thomas Reinart, Colin Shearer, and Rudiger Wirth, *CRISP–DM Step-by-Step Data Mining Guide*, 2000. http://www.crisp-dm.org.

PASW Modeler 13 Algorithms Guide. SPSS Inc, 2009. http://www.spss.com.

PASW Modeler 13 Applications Guide. SPSS Inc, 2009. http://www.spss.com.

PASW Modeler 13 Modeling Nodes. SPSS Inc, 2009. http://www.spss.com.

Rud, Olivia Parr. *Data Mining Cookbook: Modeling Data for Marketing, Risk and Customer Relationship Management*. New York: Wiley, 2001.

Tang, ZhaoHui, and Jamie MacLennan. *Data Mining with SQL Server 2005*. New York: Wiley, 2005.

Index

Page numbers in italics refer to tables and figures.

acquisition models   6, 306–7
"adults–classic social users" segment,
      mobile telephony   319–20
affinity models (*see* association models)
"affluent mass" segment, retail banking
      269, 289
"affluent" segment, retail banking   269
agglomerative models   42, 44–5, 61
"aggregate/group by" database function
      149
aggregated information   51, 160
aggregation levels   205
anomaly detection   59, 84, 100
antecedent events   51
ARPU (average revenue per unit)   217,
      296
association models   4, 7, *14*, 38, 50–56,
      64
   comparison with classification models
      56
"average shoppers" segment, retailing
      347

banking (*see* credit cards, retail banking)
Bayesian networks   25
behavioral fields   *68*, 154
behavioral segmentation   4–5, 14, 135,
      203–13
   business understanding phase   203–5
   consumer markets   192, 194–5
   credit cards   258–64
   data preparation phase   205

data understanding phase   205
   fixed telephony   329
   mobile telephony   294, *295, 316*
   retailing   334
binning (IBM SPSS Modeler)   301, *302*
"borrowers–credit card users" segment,
      retail banking   283, 284, *285, 287,*
      288, 289
brand management   214
bubble plots   242–3
business markets   200–2
"business mass" segment, retail banking
      269, 289
business meanings   73
business objectives   203–4
   retail banking   283–8
business understanding phase   11,
      203–5, 216–18
"business users" segment, mobile
      telephony   324–6

C&RT (classification and regression trees)
      algorithm   123, 125
C5.0 algorithm   124, 125
call detail records (CDRs)   163–4,
      292
"cash advance users" segment, credit
      cards   261
categorical fields   18, 21
categorization levels   227
CDRs (call detail records)   162–3,
      292

CHAID (chi-square automatic interaction detector) algorithm 125
  IBM SPSS Modeler *126, 127, 128, 129–30*
channel management 213
chi-square test 125
churn models 6, 14, 33–6, 195
classification and regression trees (C&RT) algorithm 123, 125
classification models 3, *14*, 19–32, 61
  behavioral segmentation 135
  comparison with association models 55
  decision trees 110, 111
  evaluation 25–31
  marketing applications 32–3
  scoring 32
cluster centers/centroids 101, *102*, 103, 238
cluster profiling 44, 100–8, 110, 130, 209, 223
  credit cards 237–9
  decision tree models 119–20, 209
  mobile telephony 308
  IBM SPSS Modeler 102–5
  retail banking 276–83
clustering fields 101–2
clustering models 4, 5, *14*, 39–46, 64, 82–94, 194, 195, 208, 223
  K-means 46, 83, 85–8
  Kohonen networks 46, 83, 91–4
  profiling phase 45, 100–8
  TwoStep clustering models 46, 83, 88–91
clustering solutions 96–100, 109, 127–31, 209
cohesion, clusters 97–8, 99, 100
communalities (components) 78–9
component scores 79–80, 237
confidence, decision tree rules 121
confidence measures 55
confidence scores 18, 21, 31
consequent events 51
consumer markets 191–200
continuous outcomes 37
core segments 211

mobile telephony 292–4
  retail banking 264, 268
correlation 38, 48, 66
credit cards 225–64
CRISP-DM (cross-industry standard process for data mining) process model 11
CRM (customer relationship management) 1–2, 4–8, 14
cross-selling models 6, 14, 221
current tables 163, 179
  mobile telephony 163–7
  retail banking 138–40
  retailing 179–80
customer categorization 141
customer information table
  mobile telephony 164–5, *166*
  retail banking 138, *139*
  retailing 179–81, *182*
customer profitability 217
customer relationship management (CRM) 1–2, 4–8, 14
customer segmentation (*see* segmentation)
customer signatures 35

data auditing 299–300
data enrichment 206–7
data mart (*see* mining data mart)
data mining models 3, 13
data preparation commands, IBM SPSS Modeler 162, 163, *164*
data reduction 47–50, 65, 66, 207–8, 235
data understanding and preparation phase *11*, 12, 205, 218
data validation 206
database functions 149
decision rules 24
  association models 51, 55–6
  classification models 111
  decision trees 121
  sequence models 57
decision tree models 118–19
decision trees 22–3, 110–27, 209, 211, 213, 254–6, 326–9
  advantages 121–3

C&RT algorithm   123, 125
C5.0 algorithm   124, 125
CHAID algorithm   125
   IBM SPSS Modeler   *116*, *117*, 125
deep-selling models   6, 14
deltas   155
demographic segmentation   193, 198–9
deployment phase   *11*, 12
"depositor–investor" segment, retail
   banking   283, 284, *285*, 286, *287*,
   289
"derive/compute" database function
   150
derived clusters   101
derived measures   154
   mobile telephony   177
   retail banking   155
   retailing   187
direct marketing campaigns   5–7, 14
"dormant customers" segment, credit
   cards   262–3
"dormant customers" segment, retailing
   344

eigenvalues   71–2, 73, 74
error measures
   classification models   26
   estimation models   38
estimation models   3, 18, 37–9
Euclidean distances   43
evaluation, classification models
   25–31
event outcome period (datasets)   136
"everyday shoppers" segment, retailing
   344, 345
"exceptional occasions customers"
   segment, retailing   *344*

factor analysis (*see also* PCA)   47, 66,
   206–7
"family men" segment, credit cards
   248–9, 257
"family shoppers" segments, retailing
   348
"fashion ladies" segment, credit cards
   251–2, 258

"fashion shoppers–single women"
   segment, retailing   *347*
field screening models   36–7
fixed telephony   329–31
flag fields   155
fraud detection   60–1
"full relationship customers" segment,
   retail banking   283, 284, *285*, 286,
   287, 288
"fun lovers" segment, credit cards
   252–3, 257

gains measures   28, 30
generalized rule induction (GRI)   56
Gini coefficient   123–4
"golden customers" segment, retailing
   344
GRI (generalized rule induction)   56

"heavy buyers–transactors" segment,
   credit cards   261–2
hierarchical clustering (*see* agglomerative
   models)
historical period (datasets)   135
hit propensities   27
"hobby shoppers–single men" segment,
   retailing   *347*
hypothesis testing   68

ID fields   58
"inactive mass" segment, retail banking
   269, 289
incoming network table, mobile telephony
   170, *171*
incoming usage table, mobile telephony
   169–70
industry-based segmentation   201, *202*
information gain measure   124
input fields   3
input–output mapping functions   3
"insert/append" database function   149
"installation buyers" segment, credit cards
   262
"international users" segment, mobile
   telephony   322–4

K-means   46, 83, 85–8, *131*
  comparison with Kohonen networks
    93
  IBM SPSS Modeler   84, 88–89, *91,*
    94–6
key performance indicators (*see* KPIs)
key usage attributes   182
Kohonen networks   46, 83, 91–4, *131–2*
  comparison with K-means   93
  IBM SPSS Modeler   84, *97, 98*
KPIs (key performance indicators)   102
  credit cards   229, 243, *244, 259*
  mobile telephony   305–6

latency period (datasets)   135
life-cycle segmentation (consumer
  markets)   193, 198, 199
life-stage segmentation (business markets)
  201, *202*
life-time value   217
lift measure   28, 55
Likert scales   49
linear correlation   38, 48, 66
logistics regression   35
lookup tables   138
  mobile telephony   170–2
  retail banking   143–8
  retailing   179, 183
loyalty-based segmentation   193, 196–8

machine learning   61–2
management strategy, segmentation
  213–16
market basket analysis   51
market research   209–10
marketing applications, classification
  models   32–3
marketing services management   214
MARPU (marginal average revenue per
  unit)   217, 294, 295, 296, 301, 304,
  305, 316–17
"mature users" segment, credit cards
  257
maximum benefit points   31

MCIFs (marketing customer information
  files)   148–55
  mobile telephony   172–7, 295
  IBM SPSS Modeler   150, *152*, 153
  retail banking   155–60, 271
  retailing   184–7
"merge/join" database function   149
mining data mart   133–87
  mobile telephony   160–77, 295
  retail banking   137–48, 155–60, 271
  retailing   177–87
mobile telephony
  customer profitability   *217*
  mining data mart   160–77
    current tables   163–7
    lookup tables   170–2
    MCIFs   172–7
    monthly tables   167–70
  segmentation   194–5, 197–8, 204,
    291–329, *331*
model evaluation phase   *11*, 12
modeling phase   *11*, 12
monthly tables
  mobile telephony   167–70
  retail banking   140–3
  retailing   180–3
multicollinearity (correlation)   50

naming, segments   210
NBA (next best activity strategies)   8–10
needs/attitudinal segmentation   193,
  199–200, 333
network codes table, mobile telephony
  172, *173*
network types table, mobile telephony
  172, *173*
neural networks   24–5
next best activity (NBA strategies)   8–10
null hypotheses   68

"occasional users" segment, credit cards
  253–4, 258
"older families/retired" segment, retailing
  347, 348
"oldies" segment, credit cards   250–1

"oldies–basic users" segment, mobile telephony 317–19
OLSR (ordinary least square regression) 37
"one-off/occasional buyers" segment, credit cards 263
ordinary least square regression (*see* OLSR)
outgoing network table, mobile telephony 170
outgoing usage table, mobile telephony 168–9
outliers (*see also* anomaly detection) 39, 46, 60, 90, 206

Pareto principle 221, 267, 305
IBM SPSS Modeler
   anomaly detection 60, 84, 100
   CHAID algorithm *126, 127, 128, 129–30*
   cluster profiling 102–5
   data auditing 299–300
   data preparation commands 162, 163, *164*
   decision trees *116, 117,* 125
   K-means 84, 87–8, *91*
   Kohonen networks 84, *94–96, 97, 98*
   MCIFs 150, *152,* 153
   PCA 80–2, 230
   RFM analysis *339–343*
   segmentation 230, 238, 269, *270,* 301, *302, 303, 328*
   TwoStep clustering models 62, *91*
PCA (principal component analysis) 47, 65–82, 126, 206–7
   communalities 78–9
   component scores 79–80
   credit cards 234–6
   mobile telephony *307–12*
   IBM SPSS Modeler 80–2, 233
   retail banking 274–6, *277–8, 279–80*
Pearson correlation coefficients 48, 66, 67
personalization (data) 178

"postpaid–contractual customers" segment, mobile telephony 293
potential value 217
pre-clusters 89, 90
predictive models (*see also* supervised models) 135
predictors 18, 117–8
"prepaid customers" segment, mobile telephony 293
principal component analysis (*see* PCA)
product affinities 51
product codes table, retail banking 144–5, *147*
product groups 52
product groups table
   retail banking *146*
   retailing 182–3
product hierarchy table, retailing 183, *184*
product management 214
product ownership and utilization table, retail banking 141, *143*
product status table, retail banking 138, *140*
product suggestions 51
profiling (*see* cluster profiling)
profiling phase 208–10
profitability curves 219, *220,* 268, *269*
propensity scores 3, 21, 195–6
propensity-based segmentation 192, 195
pruning, decision trees 119
"pure mass" segment, retail banking 270–1, 276, 289

rate plan codes table, mobile telephony 171
rate plan history table, mobile telephony 165, *167*
ratios 154
record screening 59–61
relative error measure 38
response measures 27, 28
"restructure" database function 150
retail banking
   customer profitability *218*

retail banking (*Continued*)
  mining data mart   137–48, 155–60
    current tables   138–40
    lookup tables   143–8
    MCIFs   155–60
    monthly tables   140–3
  segmentation   141, 194, 264–90
retailing
  mining data mart   177–87
    current tables   179–80
    lookup tables   179, 183
    MCIF   184–7
    monthly tables   180–3
  segmentation   333–48
return on investment charts (ROI charts)
    30
"revolvers" segment, credit cards
    260
RFM (recency, frequency, monetary)
    analysis   168, 334–46, 348
  IBM SPSS Modeler   339, *340,*
    341–3
RFM scores   337–8
ROI (Return On Investment) charts
    30
rotation (components)   75–6
rules (*see* decision rules)

scoring models   211
scree plots   73–4
segment and group membership table,
    retail banking   141, *142*
segment migration   222
segmentation   4–5, 14, 41, 65–132,
    189–224
  behavioral segmentation   203–13
  business markets   200–2
  cluster models   82–94, 190
  cluster profiling   100–8, 110, 130
  clustering solutions   96–100, 108–9,
    127–31
  consumer markets   191–200
  credit cards   225–63
  decision trees   110–27

fixed telephony   329–31
  management strategy   213–16
  mobile telephony   291–329, *331*
  IBM SPSS Modeler   233, 237, *270,*
    301, *302, 303, 328*
  PCA   65–82, 126
  retail banking   264–90
  retailing   333–48
segmentation criteria   204
segmentation fields
  credit cards   *230–3*
  mobile telephony   295–300
  retail banking   *272–3*
segmentation levels   204–5, 227
segmentation populations   204, 228–30
segmentation types   191, *192,* 223
"select" database function   149
self-organizing maps (SOMs)   46, *92,*
    93–4
separation, clusters   98–100
sequence models   4, 8, *14,* 56–9
service types table, mobile telephony
    171–2
"share of wallet"   266
size-based segmentation   201
socio-demographic segmentation (*see*
    demographic segmentation)
SOMs (self-organizing maps)   46, *92,*
    93–4
SQL   149, 150, 162, 163
standard deviation   72, 95
standardized fields   73
statistical techniques   61–2
sum of squares between (SSB)   99
sum of squares error (SSE)   96
"superstars" segment, retailing   344
supervised models (*see also* classification
    models, estimation models, field
    screening models, voluntary churn
    models)   3, 17–39, 61–2
  cluster profiling   110
  segmentation   254–6
supervised segmentation   110
support, decision tree rules   121

support measure   54
SVM (support vector machine)   25

"telcos fans" segment, credit cards
    247–8, 257
telecommunications (*see* fixed telephony,
    mobile telephony)
terminal nodes, decision trees   114
top-up information, mobile telephony
    169–70
training phase   17, 135, 212
transaction channels table, retail banking
    145, *146*, *147*
transaction codes table, retail banking
    145, *147*
transaction data   52, 53–4
    retail banking   141–8
    retailing   180, *181*
transaction types table, retail banking
    145, *148*
transactions table
    retail banking   142, *144*
    retailing   180, *182*
"travelers" segment, credit cards
    249–50, 257
TwoStep clustering models   46, 83,
    88–90, *131*
    IBM SPSS Modeler   62, *91*
"typical customers" segment, retailing
    344
"typical users" segment, credit cards
    246–7, 256

"unclassified" segment   213
unsupervised models (*see also* association
    models, clustering models, sequence
    models)   3–4, 39–61, 63–4
    data reduction   47–50
    record screening   59–61
unsupervised segmentation   127
up-selling models   6, 14, 221
"update" database function   150

validation phase   136
value-at-risk segmentation   14
value-based segmentation   14, 216–22,
    223
    business markets   200–1
    consumer markets   191, 193–4
    mobile telephony   294, 296, 300–7,
        316
    retail banking   267–8
    retailing   333
variance   72–3
Varimax rotation (PCA)   75–6, 234
voluntary attrition models   6
voluntary churn models (*see* churn
    models)

"young –SMS users" segment, mobile
    telephony   321–2
"young active customers" segment, retail
    banking   283, *284*, *285*, 286, 287,
    289